THE GIZA DEATH STAR REVISITED

An Updated Revision of the Weapon Hypothesis of the Great Pyramid

JOSEPH P. FARRELL

Adventures Unlimited Press

Other Books by Joseph P. Farrell:

Hess and the Penguins
Hidden Finance
The Third Way
Nazi International
Thrice Great Hermetica and the Janus Age
Covert Wars and the Clash of Civilizations
Saucers, Swastikas and Psyops
Covert Wars and Breakaway Civilizations
LBJ and the Conspiracy to Kill Kennedy
Roswell and the Reich
Reich of the Black Sun
The S.S. Brotherhood of the Bell
Babylon's Banksters
The Philosopher's Stone
The Vipers of Venice
Secrets of the Unified Field
The Cosmic War
The Giza Death Star
The Giza Death Star Deployed
The Giza Death Star Destroyed
Transhumanism (with Scott deHart)
The Grid of the Gods (with Scott deHart)

THE GIZA DEATH STAR REVISITED

An Updated Revision of the Weapon Hypothesis of the Great Pyramid

The Giza Death Star Revisited

by Joseph P. Farrell

ISBN:978-1-948803-57-1

Published by:
Adventures Unlimited Press
One Adventure Place
Kempton, Illinois 60946 USA

auphq@frontiernet.net

www.adventuresunlimitedpress.com

Cover by Joe Boyer

10 9 8 7 6 5 4 3 2

Table of Contents

To all my friends
and to Catherine Austin Fitts and all the donors to the virtual pipe organ crowd fund;
and all the "Gizars" who have stimulated and entertained speculative inquiry and conversation, some of the fruits of which are found herein;
and to G.A.H., and
Especially to T.S.F. who encouraged me to write down all my "strange ideas" about the Pyramid:
you will always be profoundly missed;
and to my sweet little dog Shiloh who defines the essence of loyalty, love, and friendship.

"Thank you" is inadequate,
but I hope in some measure this book manages to say it.

Preface, Or Why I decided to (re)Write This Book

"Sir, we're still all agreed your theory was, and remains, crazy. The question still dividing us remains whether it is crazy enough."
Nils Bohr to Werner Heisenberg[1]

It seems like a dream now, but a little over twenty years ago, during a period of major financial difficulty and upheaval in my life, and at the prompting of (1) a friend, of (2) a brother-in-law, and of (3) a professional and professorial colleague, I was encouraged to write a book that would become my first foray into the world of "alternative research." That book was *The Giza Death Star*, and little did I know at the time it was published, it would soon launch a career in alternative research book-writing, conference appearances, and even radio and television interviews that would be successful beyond my wildest dreams, and become a fulltime career and livelihood. It was a career I certainly never *intended*, but which I certainly had and have no objections to. The only thing I intended when I wrote the book was that if it was even remotely successful, I had a few "follow-up" books in mind.

As I said, to my great shock and surprise, the book was "successful," at least, as successful as such books can be. My one regret was that when writing that book and its two sequels, *The Giza Death Star Deployed* and *The Giza Death Star Destroyed*, I was new to the field of alternative research, and in fact, I was still hoping that I would be able to have some sort of academic teaching career. Accordingly, I wrote the books in such a way that I "pulled my punches" and over-simplified or conversely "over-

[1] The original version of this quotation appeared as the epigraph to the Preface of *The Giza Death Star*. By modifying the quotation as I have, I mean to indicate that the weapon hypothesis remains no less crazy now than when I first advanced it, and indeed, though I'm not really the origin of the hypothesis, I've nevertheless fleshed it out far beyond its original proposer did.

complexified" when I should not have. I avoided disclosing my most outlandish thoughts about what was already an outlandish hypothesis. Over the years since then I have thus expressed on many occasions regret about the way the books were written to the extent that, had I to do them all over again, I would have written those three books in a profoundly different way. When their publisher—David Hatcher Childress of Adventures Unlimited Press—informed me that he had allowed the books to go out of print and that their copyright had reverted to me, I thought "now is my opportunity" to go back and redo the whole series, but after consideration, I have decided that this would do a disservice to the record, and that the books should be maintained in print in their original form, and that a fourth book should be done which would allow me to "revise and extend" my remarks.

My purpose in re-writing and condensing the original *Giza Death Star* trilogy into this one book is twofold:

1) Firstly, to give an updated, revised, and distilled exposition of the *core* of the original argument for the weapon hypothesis that may have been unclear due to its being dispersed among the original three books. Moreover, the two foundations of that argument may also have been unclear.
   a) The *first* foundation consisted and consists of the contextual indications alluding to the function of the Great Pyramid, or pyramids in general, as a weapon in various texts; and,
   b) The *second* foundation consists of the "hardware" or structure itself, i.e., those elements of the building, including hypothesized missing hardware, suggesting a weapon function and purpose for the structure.

   As an undapted, revised, and distilled presentation of the core argument of the original trilogy, I intend this book to be a kind of "study guide" for those readers possessing the original trilogy. As such, none of this book is to be taken as negating the importance of material presented in those books which is *not* reprised here. It merely means that much of that material does not constitute the core argument.

2) Secondly, I also intend this book to be a one-volume stand-alone presentation of the weapon hypothesis argument for readers unfamiliar with the original trilogy.

The problem in writing such a book is that whole sections of the original trilogy are repeated here, though in a different order and context to highlight the argument and hopefully make it clearer. When doing so, I have noted such sections with brackets — "[" and "]" to indicate the beginning and end of such sections. These sections will include all original quotations, citations, and footnotes from the original trilogy, though obviously their numeration will have changed. After the concluding bracket, "]", a new footnote is included referencing the original book from the trilogy from which the passage is repeated.

As the reader of this book might expect, in these instances, the excerpted portions can be quite long, and additional new or explanatory comments in those sections which I have added in *this* book are simply noted by placing them inside of **(boldface parentheses like this),** rather than breaking the flow with more brackets and footnotes.

A final note and caveat must be mentioned. Even though approximately half of this book consists of repeated material from the original trilogy, about half consists of new material, or added explanatory material. Thus, the reader of the original trilogy is advised *not* to skip through sections that "he thinks he knows," because the new order or arrangement of the argument, the expanded textual context from my other books, and the addition of new materials, has made for an enhanced and clearer argument. In particular, three chapters in this book constitute the "core of the core" argument:

1) *Chapter 1* outlines the detailed core of the *hardware component* of the argument, and the wider analogical context for it in Napoleon's expedition, and the uncanny

qualitative resemblance of its etchings of Giza to modern military phased array radar compounds;

2) *Chapter 3* surveys the various texts bearing directly on the weapons function of the Great Pyramid, including Sitchin's original exposition of the weapon hypothesis, and those texts mentioning the Tower of Babel story in connection to a threat to God or the gods;
3) *Chapter 7* details the electrical impulse technologies of Tesla and T.T. Brown, the dimensional analogues of geodetic and celestial space in the structure and their functional purpose, and the recent metamaterials-crystal technologies that tend to confirm my original hypothesis of "φ" or "phi" crystals I speculated made the structure work.

In my original introduction to the first book of the original trilogy, I mentioned that I had never felt anything "good" about the Great Pyramid, that it did not fill me with "positive vibes" as it seemed to do for so many people. I have not changed my view on that score, nor on the weapon hypothesis, one iota. I remain awed by the structure, and by whomever built it, but for very different reasons having little to do with the presumed jonquils and daisies of a bygone "golden age." There was no such thing; the texts are clear, and fairly universal, in presenting that bygone age as an age of rebellion, of stupendous wars fought with colossal weapons leading to abysses of destruction.

So if you're looking for warm fuzzies and blankets and jonquils and daisies, this is probably not the book for you.

With these things in mind, it's time to revisit *The Giza Death Star* trilogy, and to repeat, revise, and extend our original remarks.

Joseph P. Farrell
From Somewhere, 2023

PART ONE:
TEXTUAL AND CONTEXTUAL EVIDENCE

"It is reasonable to assume that if we were to destroy ourselves through nuclear holocaust, the geological and biological record would bear witness to it, and reveal that knowledge to future archaeologists as they became more advanced in their science. At the same time, some of our civil engineering projects might survive, and the occasional archaeological anomaly might turn up to promote some thought in that direction."
Christopher Dunn, *The Giza Powerplant*, p. 244.

1.
A Remarkable and Remarkably Strange Structure, with some Remarkable Claims

*"... there is no record in Egypt itself of any gradual development of architectural knowledge and skill. How did the exquisite technical knowledge and skill displayed in this vast structure suddenly make its appearance in this mysterious land? We might ask the further question, which, indeed, has been asked before. Though the Great Pyramid is **in** Egypt, is it **of** Egypt?"*
William Kingsland, *The Great Pyramid in Fact and Theory*[1]

"I have devoted much of my time during the past year to the perfecting of a new small and compact apparatus by which energy in considerable amounts can now be flashed through interstellar space to any distance without the slightest dispersion."
Nichola Tesla, *The New York Times*, Sunday, 11 July, 1937.[2]

A. Alternative Views

DEFIANCE: IF THERE IS ONE WORD THAT DESCRIBES the Great Pyramid of Giza, it is defiance, for it stands at the edge of the Libyan desert, but also at the edge of the fertile Nile flood plain, mutely towering over its surroundings and seemingly defying all attempts to explain what it was or what it is or what it was built to be and designed do and how it did it. For some people who call themselves by the grand name of "Egyptologist," it was designed to be an elaborate sepulchral tomb, though admittedly, no corpses—mummies or otherwise—were ever discovered whiling

[1] William Kingsland, *The Great Pyramid in Fact and Theory* (Literary Licensing reprint, ISBN 9781497971417), p. 10, emphasis in the original.

[2] This appeared as the original epigraph to chapter five of *The Giza Death Star Deployed*, p. 141.

away their eternity inside the structure.[3] For others, it wasn't even designed to do *that*. Rather, they maintain that it was deliberately conceived and designed to do nothing but squat there and simply "be", in all its lithographic glory, a permanent record of all sorts of arcane dimensional analogues of the cosmos at large, a very large and expensive bureau of standards, weights, and measures.[4] For still others, it was a "miraculous" structure, based on systems of measure known only to its "divinely inspired" and presumed Hebrew builders but unknown to their "profane Egyptian" masters; units of measure which, moreover, were conveniently close to British Imperial measures and which allowed the whole structure to be "decoded" as a kind of "biblical prophecy in stone", and a key to the entire history of the human race, past, present, and future.[5] And for yet others, this whole notion of a prophecy in

---

[3] For a good one-volume standard "Egyptological" introduction to the Egyptian pyramids as a whole, considered in relation to the Egyptian dynasties, see Miroslav Verner, *The Pyramids: The Mystery, Culture, and Science of Egypt's Great Monuments* (New York: Grove Press, 2001, ISBN 0-8021-3935-3), trans. from the German by Steven Rendall.

[4] For the best articulations of the "bureau of weights, standards, and measures," theory, see John Taylor, *The Great Pyramid: Why was it Built? And who Built it?* (London: Longman, Green, Longman, and Roberts, 1959; Reprinted: Cambridge University Press, 2014, ISBN 978-1-108-07578-7), especially, pp. 122-126; see also William Kingsland, *The Great Pyramid in Fact and Theory* (Literary Licensing, ISBN 9781497971417).

[5] The standard view of the Pyramid as a prophecy in stone remains Piazzi Smyth's *Our Inheritance in the Great Pyramid*. A more recent, and succinct, exposition of this view from a typical "Baptist" evangelical view is N.W. Hutchings *The Great Pyramid: Prophecy in Stone* (2010: Noah Hutchings, ISBN 0-98362164-0). Another book seeking to correlate the astrological "Gospel in the stars" interpretation with the Pyramid as a Prophecy in Stone is E. Raymond Capt, M.A., A.I.A., F.S.A. Scot, *Study in Pyramidology* (Artisan Sales, 1986, ISBN 0-934666-20-2). A broader and somewhat less dispensational view is found in Peter Lemesurier, *The Great Pyramid Decoded* (New York: Avon Books, 1977, ISBN 0-380-43034-7). Then there's the whole "Pyramid as prophecy" combined with Edgar Cayce school, exemplified by John Van Auken's *2038: The Great Pyramid Timeline Prophecy* (Virginian Beach, Virginia: 2018: 4th Dimension Press, 978-0-87604-699-9.

stone based on units of measure close to British Imperial was a putrid pile of rocky nonsense, notwithstanding the fact that they for whom this *was* all nonsense nonetheless wanted to rob the prophecy theorists of some of their metrological insights in order to confect a golden calf of theoretical speculation of their own.[6] Still others who are willing to follow the clues of engineering rather than the fantasies of Egyptology or evangelical dispensationalism have thought it to be a power plant or machine of some sort,[7] and still others have argued that it was an "alchemical" machine for the manufacture of monatomic gold.[8]

And lest it be forgotten, the Soviet Union during the 1970s and 1980s was conducting top secret black projects research on pyramids and came to some intriguing conclusions, some of which will be reviewed in this book. Not to be left behind in a "pyramid gap", persons closely associated with the US government and with the whole mystery of vanishing civilizations and UFOs also investigated the structure.[9] Over all this debate, the Great Pyramid sits silently, defying attempts even to *measure* it accurately and to explain how in the name of sense a bunch of Egyptians (or Hebrews, or whomever) managed to build the mysterious monument with near optical precision using copper saws, ramps, pullies, and, as the classical historian Herodotus reported, lots of onions and garlic to feed tens of thousands of workers whom, he also noted, somehow built the structure from the top down! We'll return to Herodotus in the course of this book, but for now, the key word for the structure is *defiance*.

It is a monument *of* defiance and *to* defiance, for no matter what theory one advances to explain it, the structure at some point

---

[6] The best, and earliest refutation of the "prophecy in stone" view of the structure still remains William Kingsland's *The Great Pyramid in Fact and Theory* (Literary Licensing reprint, ISBN 9781497971417).

[7] Christopher Dunn, *The Giza Power Plant: Technologies of Ancient Egypt* (1998: Santa Fe: Bear and Company, Inc., ISBN 1-879181-50-9).

[8] Spencer L. Cross, *The Great Pyramid: A Factory for Mono-Atomic Gold* (20103: Spencer L. Cross, ISBN 978-0615919768).

[9] Q.v. Nick Redfern's *The Pyramids and the Pentagon: The Government's Top Secret Pursuit of Mystical Relics, Ancient Astronauts, and Lost Civilizations* (2012: Pompton Plains, New Jersey, ISBN 978-1-60163-206-7), see especially pp. 73-84, 103-170.

defies explanation. This is no less true for the theory that I advance, but with one pointed exception. If, as I argue, the Great Pyramid was intended, built, and used as a tremendously powerful weapon, then defiance was at its very heart and core, and I argue that its strange analogues to time and space and even its stubborn refusals to be measured accurately are all potential clues to its nature and function in that capacity.

In short, I argue that it is as a monument of defiance – as a *weapon* – that the structure makes sense of the widest possible dataset, and *no other* theory covers as wide a dataset. Because of this, we shall discover that there *are* elements of each of the other theories that must be seriously considered as components of the weapon hypothesis and as possible clues to the Great Pyramid's function as such.

B. A Brief Tour of the Great Pyramid and the Metrological Problem: Its Stubborn Refusal to Be Measured Accurately

An aerial view of the Giza compound affords the best place to begin our tour.

An Aerial View of Cheops, or Khufu (the Great Pyramid) in front, Cephren in the middle, and Menkaure in the back. Note the faintly visible indentation of the visible faces of the Great Pyramid along the apothem (the line from the apex to the ground bisecting the fact, giving the faces of the Great Pyramid a "parabolic dish" character

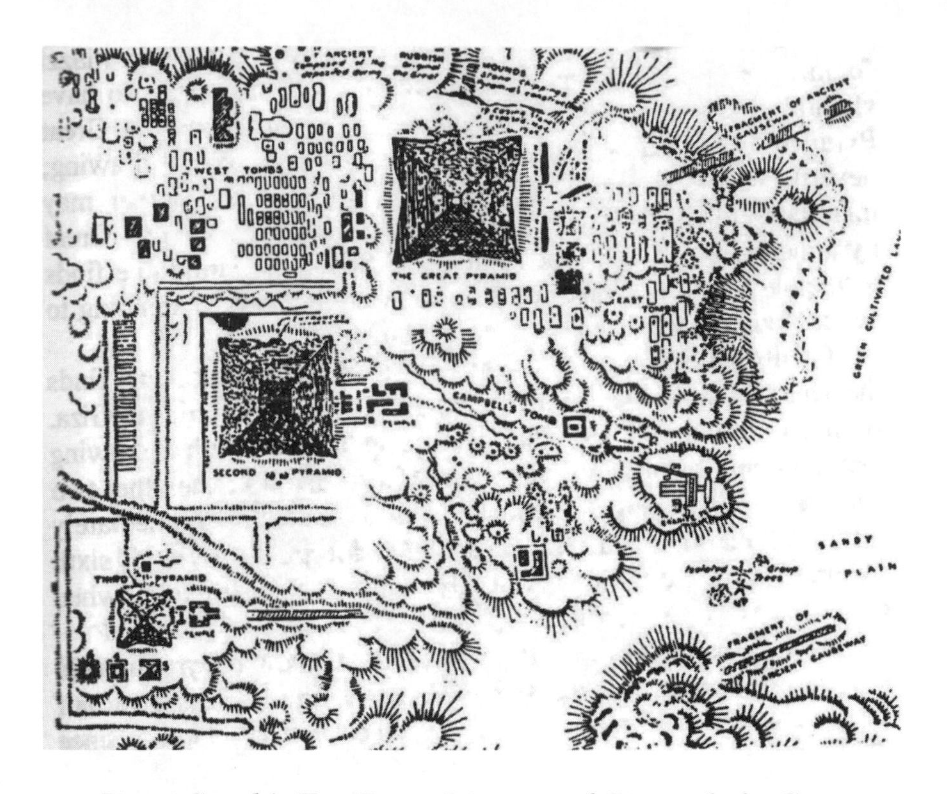

Piazzi-Smyth's Top-Down Diagram of Giza with the Great Pyramid at the top (north), Cephren in the middle and to the left, and Menkaure in the lower left hand corner. The indentation of "parabolic faces" along the apothem of the Great Pyramid is clearly visible.[10]

To the top right, one sees the Great Pyramid, its "parabolic" faces clearly in evidence. Notice the fragment of what archaeologists call a "causeway" leading away to the east-northeast. Note that if one extends the "causeway" to the Great Pyramid itself, it

[10] This diagram appears in E. Raymond Capt, M.A., A.I.A., F.S.A. Scot., *Study in Pyramidology* (Thousand Oaks, California: Artisan Sales, 1986, ISBN 0-934666-20-2), p. 18. It should be noted that Capt adheres to the "Pyramid-as-biblical-prophecy" hypothesis and to a basic dispensationalist eschatology. Indeed, he is an adherent to the "modified" Pyramid prophecy theory of J.F. Rutherford, one of the founders of the Jehovah's Witnesses. Capt's book is valuable nonetheless because of its many carefully drawn diagrams, and surveys of measurements from Piazzi-Smyth and Sir Flinders Petrie.

appears to intersect at the center or just to the south of the center of the eastern "parabolic" face.

Below and to the left of the Great Pyramid we find the Second Pyramid **(Cephren)**, the other dominant structure at Giza. Notably, in Piazzi Smyth's drawing which is reproduced here, it too appears to have slightly parabolic faces, though not as pronounced as the Great Pyramid. This, however, appears to be simply an artifact of the drawing, as Sir Flinders Petrie makes no mention of an actual indentation. Notice that to the right of the Second Pyramid one finds a "temple", with the traces of yet another "causeway" extending to the east southeast, past the Sphinx, to **(another)** granite "temple."

Finally, in the lower left hand corner of the diagram, one finds the Third Pyramid, the smallest of the three large pyramids at Giza. Again, one notes the peculiar feature in Piazzi Smyth's drawing that it too appears to have "parabolic" faces, and, like the other two pyramids, a "causeway" leading almost due east. Immediately south of the Third Pyramid are the fourth, fifth, and sixth pyramids, structures of evidently inferior construction when compared with the first three. **(Indeed, one notices, when comparing the three large pyramids, an evident decline from the highest standards of construction in the Great Pyramid, to a slightly less high standard in the Second Pyramid, and an even further degradation of standards in the Third Pyramid, a fact which as we shall see subsequently led Alan Alford to posit three distinct levels of construction at Giza, beginning with the oldest and highest standard in the Great Pyramid, a second layer of slightly declined construction much later with the Second Pyramid, the temples, and the Sphinx, and a third layer even later represented by the Third Pyramid and the rest of the compound. Much more on this subject will be stated later.)**

Now let us note one feature unique to the Great Pyramid. In thousands of years, this massive building has settled less than half an inch, in spite of numerous earthquakes in the region. Ever since Petrie's comprehensive survey of Giza, engineers have known why: beneath the Great Pyramid there are five massive stones or "sockets", four at each corner of the structure, and a fifth on the diagonal above the southeast corner. (See the figure below) These sockets are a "ball and socket" joint familiar to modern engineering, permitting the building to rock and shift gently when the Earth moves. This is the surest evidence, in and of itself, that the Great Pyramid is a coupled oscillator, for this feature is analogous to pressing down a key of a piano silently while striking another key to make its strings resonate silently. Analogously, the Pyramid, in short, was *designed to move* ***(in sympathetic vibration with the Earth at all times***).

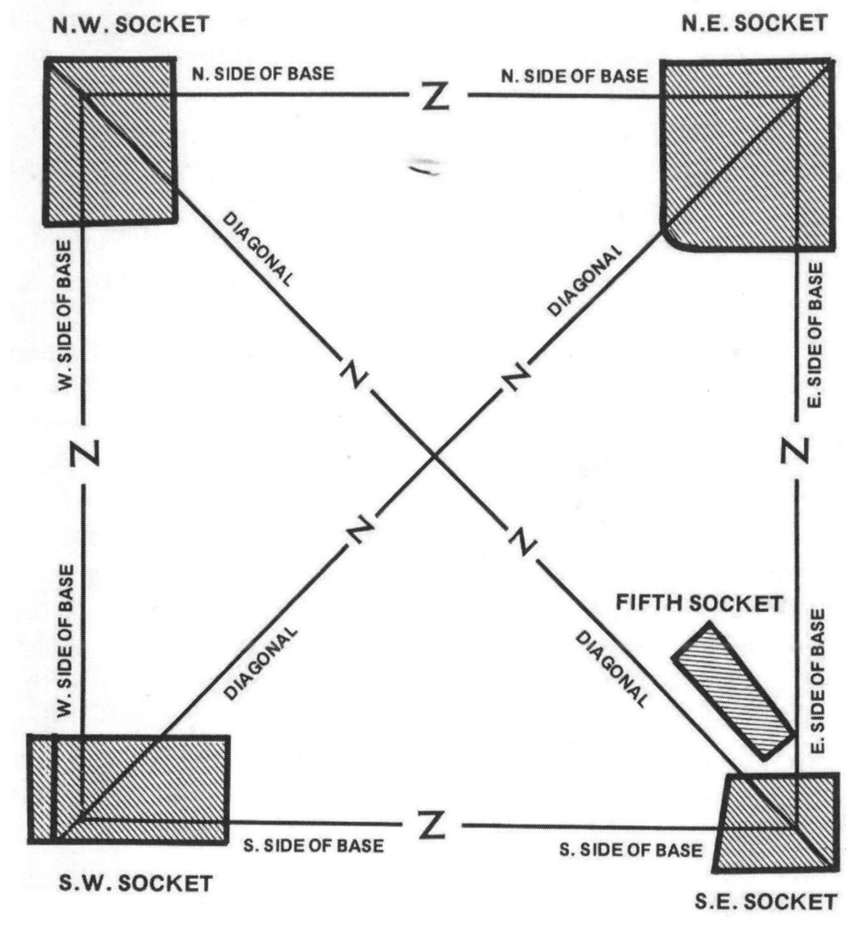

The Five Sockets at the Base of the Great Pyramid[11]

[11] Capt, *Study in Pyramidology*, p. 48.

If one now looks at a North-South cross section of the Great Pyramid, one is immediately confronted by an anomaly, as far as the pyramids at Giza go, for alone of all the pyramids there, the Great Pyramid has internal chambers **(high)** *above* the ground line, in addition to a subterranean chamber.

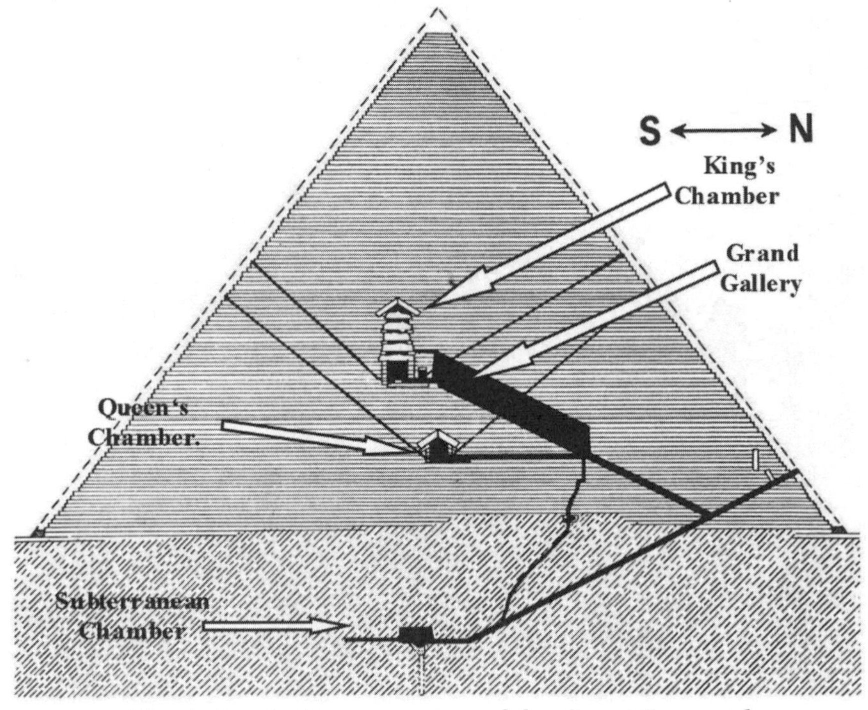

North-South Cross-section of the Great Pyramid

To the left of the axis running from the apex of the Pyramid to its base, one finds a large chamber, capped by five layers of huge roughly hewn granite stones beneath a corbeled roof. This is the so-called "King's Chamber." From this chamber two thin shafts emerge diagonally upward to the north and south faces of the structure. These are called "air shafts". Immediately to the right of the King's Chamber is a smaller chamber, called the "Antechamber:, and then, extending diagonally downward, a tall and narrow "Grand Gallery", which ends in a shaft intersecting with another shaft leading below ground to the "Subterranean Chamber."[12]

Recent research utilizing cosmic rays as a kind of radar tomography has also shown strong indication of the presence of yet another undiscovered chamber above the Grand Gallery, and perhaps of the same shape as the Grand Gallery, as the following cut-away shows:

Cutaway view of the Great Pyramid showing approximate location of presumed new chamber above the Grand Gallery

[At the lowest point of the Grand Gallery, a straight passageway extends to the lower chamber, the "Queen's Chamber", a lower and less massive chamber than the King's Chamber, but which likewise has a corbeled roof. Notice that the apex of the Queen's Chamber roof, and the upper end of the Grand Gallery—the so-called "Giant Step"—all lie on the axis running through the center of the structure up to the apex of the Pyramid itself. From the Queen's Chamber two more "air shafts" make their

---

[12] Joseph P. Farrell, *The Giza Death Star: The Paleophysics of the Great Pyramid and the Military Complex at Giza* (Kempton, Illinois: Adventures Unlimited Press, 2001, ISBN 0-932813-38-0), pp. 161-164

way diagonally upward to the north and south faces of the Pyramid, but do *not* actually emerge on the surfaces of the faces, but stop just short of it, **(thus making the idea that they are "air shafts" untenable).** Note also the layers or stone courses of the Pyramid, an important feature that we will discuss in more detail in a subsequent chapter.

Now let us look a little more closely at each of these chambers, beginning with the Grand Gallery. In terms of its sheer size, this is the largest of the interior chambers of the Great Pyramid, and it has a number of unusual features. (See the diagrams below)

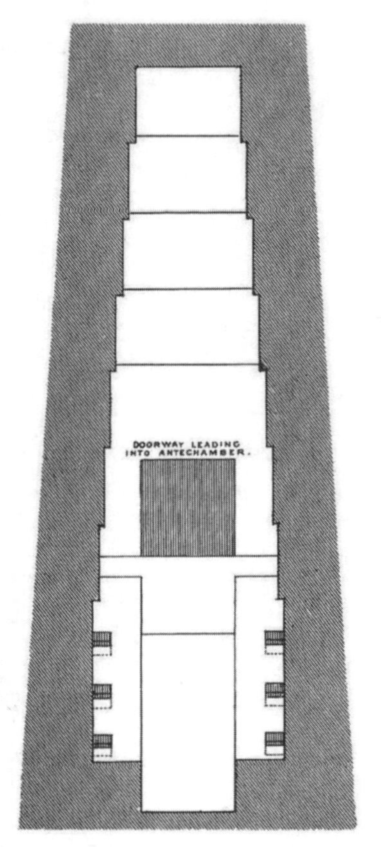

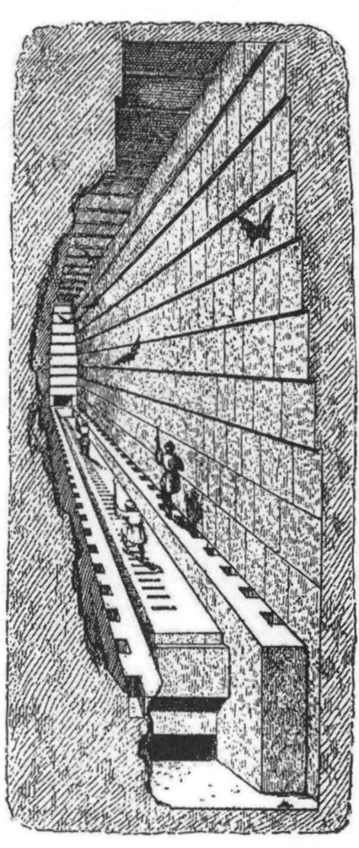

Cross-Section and Perspective Views of the Grand Gallery, with human figures to scale[13]

[13] Capt, *Study in Pyramidology*, p. 79.

Along each side of the Gallery, there is a narrow flat section, into which twenty-seven notches are cut at equal distances on each side. Not only this, but the Grand Gallery's walls narrow from the bottom to the top **(in discrete increments, rather than as a smooth process. This is a significant clue as will eventually be seen. Furthermore,)** the stones of the roof of the Grand Gallery are angled **(towards the King's Chamber).**

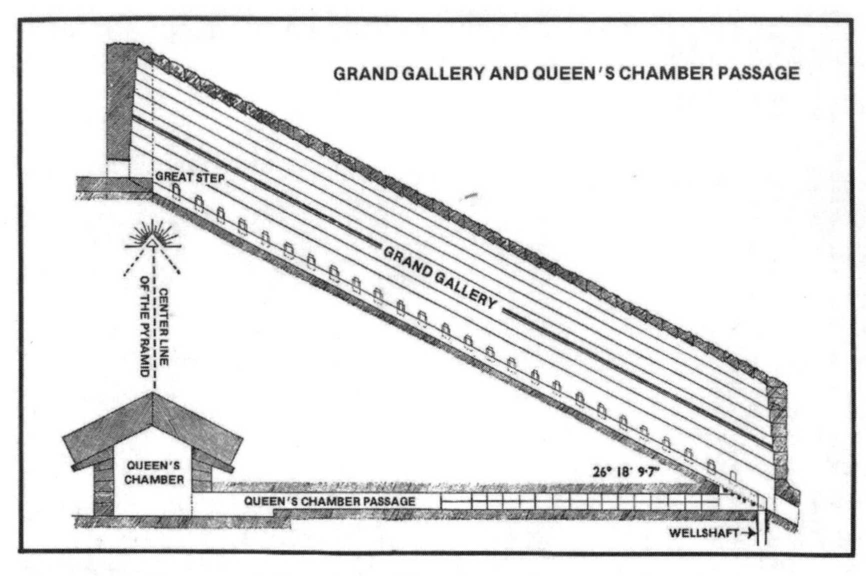

Grand Gallery and Queen's Chamber, showing the sloping stones of the Grand Gallery roof. The dotted line running through the apex of the Queen's Chamber is the centerline of the Great Pyramid.[14]

At the **(upper)** end of the Grand Gallery there is a "low passage" leading to the "Antechamber", followed by another low passage leading into the King's Chamber (see below). Inside the King's Chamber one finds a large oblong granite box, one corner of which looks as if it has been melted, called the "Coffer." Note the "air shafts" leading up from the King's Chamber.

---

[14] Capt, *Study in Pyramidology*, p. 82.

The End of the Grand Gallery, the first low passageway, the Antechamber, the second low passageway, the Coffer, and the King's Chamber[15]

Before we enter the King's Chamber, there are a number of peculiar things we must observe in the Antechamber. First, look at Haberman's perspective view of the Antechamber:

[15] Capt, *Study in Pyramidology*, p. 89.

Haberman's Persepctive Side View of the Antechamber[16]

Notice that immediately after you emerge from the first low passage, there is a large slab **(of granite)** with what appears to be a protrudence resembling a horseshoe **(or "boss")** in the center.

[16] Capt, *Study in Pyramidology*, p. 93.

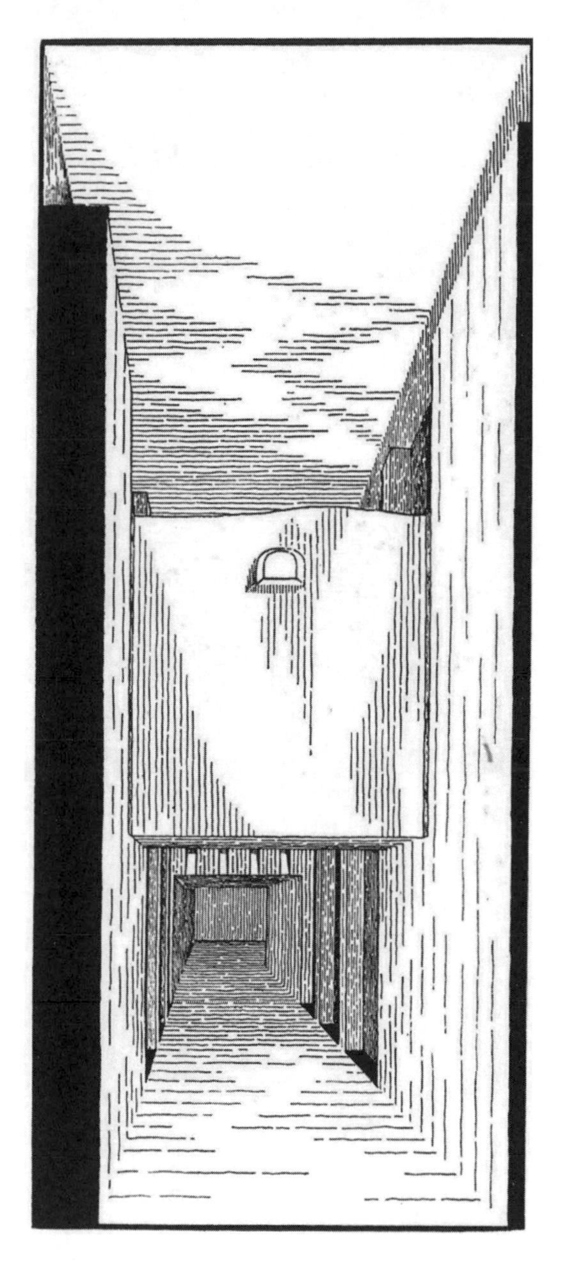

Front Perspective View of the Antechamber, showing the granite leaf and protruding "horseshoe" or "boss".[17]

[17] Capt, *Study in Pyramidology*, p. 91.

If one were to remove this slab, one could see the rest of the Antechamber:

The Antechamber with the Granite Leaf removed.[18]

---

[18] Capt, *Study in Pyramidology*, p. 90.

Looking again at **(Haberman's Perspective Side View)**, you will also note, on each side of the Antechamber behind the slab, three large slots, at the top of which are three semi-circular notches. This feature has led some to speculate that at one time there were three movable slabs of rock set into these notches which could be raised or lowered rather like a portcullis. As we shall see, Christopher Dunn has a most ingenious explanation of what purpose these notches and the "portcullis" may once have served.

Haberman's Side Perspectice View of the Antechamber once again, showing the semi-circular slots for a "portcullis".[19]

[19] Capt, *Study in Pyramidology*, p. 93.

Continuing our journey into the King's Chamber itself, we find the Coffer with its "melted" corner lying toward the western wall of the Chamber (see below and the next page):

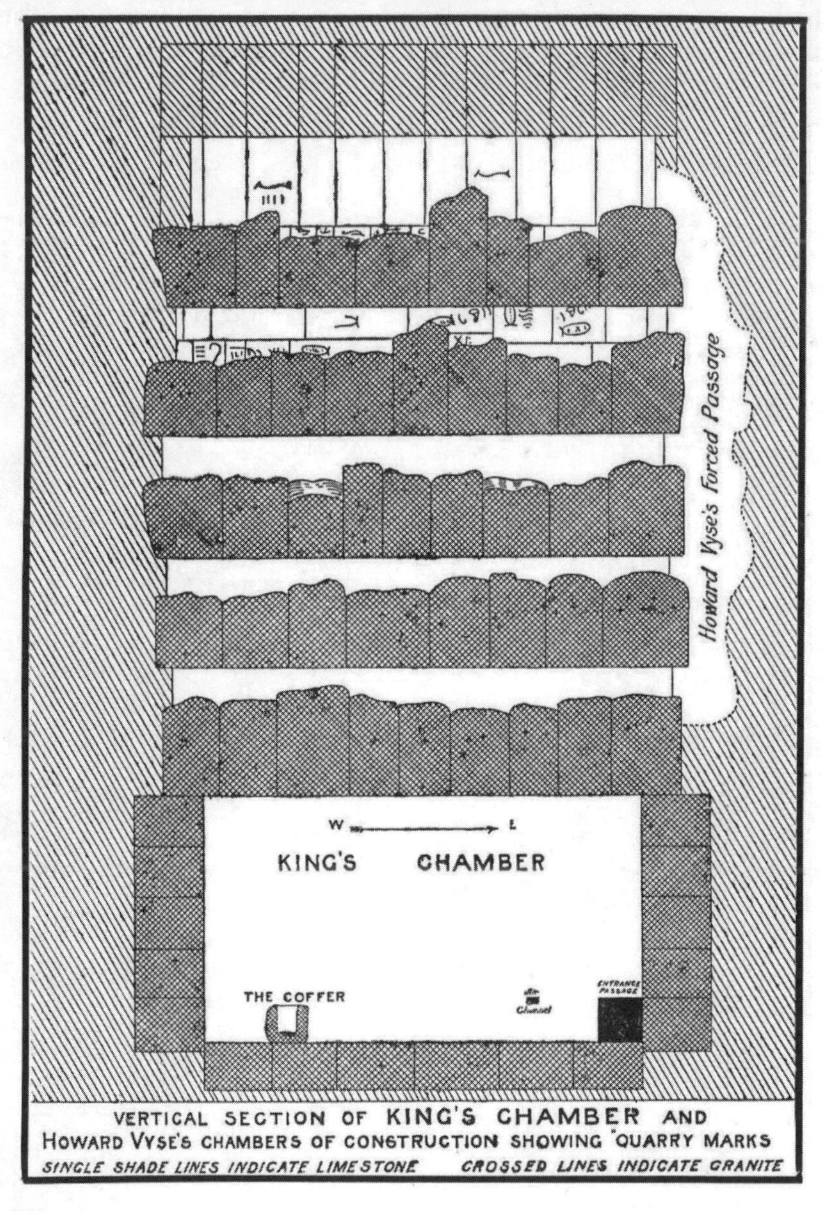

King's Chamber looking north, showing approximate location of the Coffer; note the low passageway entrance on the lower right corner, and Colonel Vyse's forced passage on the upper right.

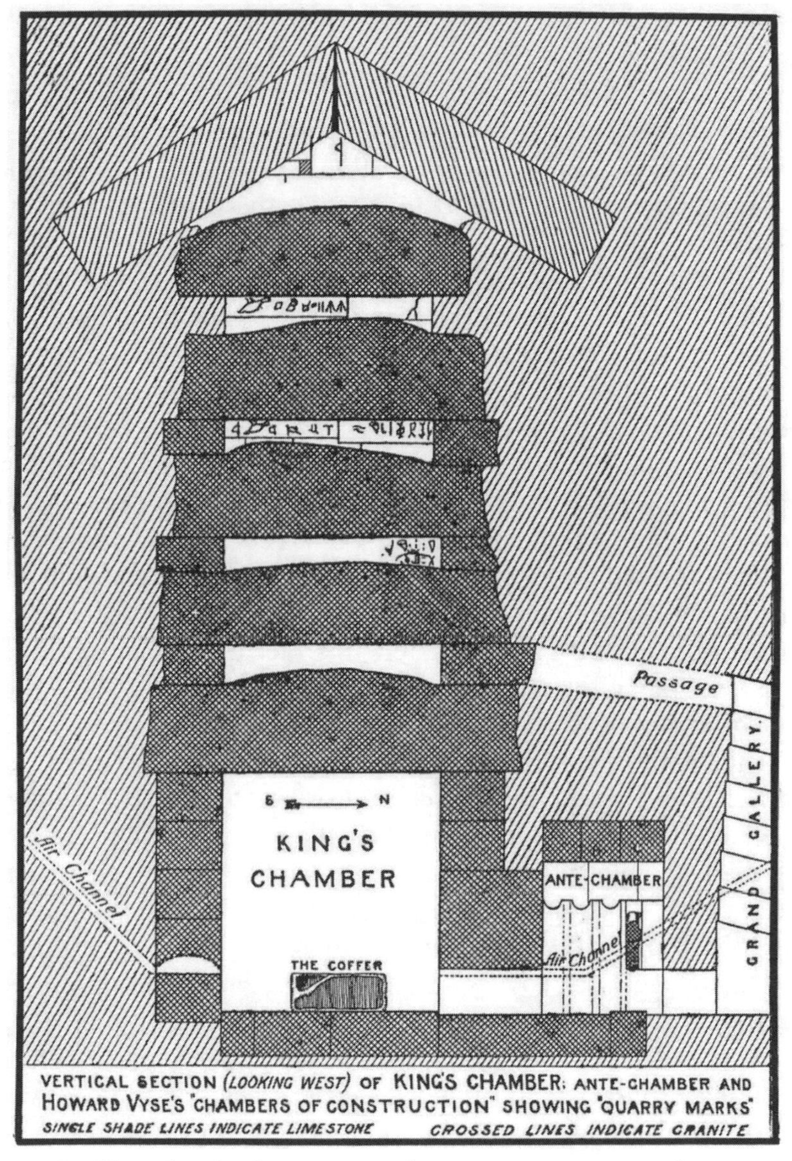

King's Chamber looking west, showing approximate location of the Coffer.[20]

Looking at **(these two cross-sections of the King's Chamber)**, we discover five layers of very large granite slabs with flat bottoms

[20] Capt, *Study in Pyramidology*, p. 94 for this diagram; the diagram on the previous page is found in Capt, p. 96.

but very roughly carved **(and unfinished)** tops, yet another seeming anomaly in a structure so perfectly constructed. Looking **(at these diagrams**) for a moment longer, we discover in the three uppermost chambers *the only hieroglyphics ever found inside the Great Pyramid*, and found in a most unlikely place, namely, *not* in the Chamber that is supposed to contain the actual sarcophagus, the Coffer.][21]

In the original *Giza Death Star*, I went on at this point to note that the British amateur archaeologist, Colonel Howard Vyse, claimed to have "discovered" these hieroglyphics (along with the chambers they were "discovered" in), much to the relief of orthodox Egyptology, for the hieroglyphs contained the cartouche of the Pharaoh Cheops, or Khufu, the alleged builder of the Great Pyramid according to Egyptology. The problem was (and remains) that the hieroglyphs were themselves of a very illiterate nature, and not what one would expect from such a lavish expenditure for a tomb, nor what one would expect from a structure otherwise so perfect. As we shall discover later, this fact led Zechariah Sitchin and others, including Alan Alford, to initially question the validity and legitimacy of Vyse' discovery and to speculate that Vyse in fact forged the hieroglyphs himself in order to lay claim to a significant discovery, and perhaps in order to provide "evidence" that the Great Pyramid is an Egyptian construction, and not something far older. For reasons that will be examined in detail subsequently—including the recently uncovered evidence of Scott Creighton proving Vyse forged the hieroglyphs—I adhere to the view that Vyse's "discovery" is a complete hoax. The importance of the Vyse question cannot be over-estimated, for his alleged discovery of the hieroglyphics is the *only* solid evidence Egyptology can point to connecting the Great Pyramid to Egypt and to the idea that it was intended as a sepulcher. If the Vyse "discovery" falls, then alternative explanations of the structure must be considered far more seriously.

Continuing with our tour, if one examines the cross-section diagrams of the King's Chamber, [one notes that the large granite stones comprising the roofs of the five chambers above the King's Chamber are all vertically cantilevered. One notes also the

[21] Joseph P. Farrell, *The Giza Death Star*, pp. 164-167.

passageway that Vyse forced up through the stone courses to reach these chambers. Once on the inside, he reported that he became covered with a fine powdery black soot. Finally, note where the low passageway emerges in the King's Chamber. Just to the left one will see the entrance of the "air shaft" into the chamber, and on the south wall directly opposite, the other shaft emerges into the chamber.[22]

C. Napoleon in Egypt: The Strange Resemblance between an 18th Century French Etching, and a Modern Military Installation

The story of Napoleon Bonaparte's strange expedition to Egypt is well-known and has been told often enough elsewhere. Here we only observe that in addition to several thousands of French troops and a large French fleet, Bonaparte also took with him the crème de la crème of France's intellects to study Egypt, and in particular, the pyramid complex at Giza. The Mamaluke troops of the Ottoman Bey (king) of Egypt, though grossly outnumbering the French, were no match for Napoleon's superbly trained and equipped soldiers, not to mention Napoleon's tactical genius, and were defeated at the battle of the Pyramids, in what may be in some sense regarded as a continuation of Zechariah Sitchin's ancient "pyramid wars."

Napoleon Bonaparte being acclaimed after the French victory at the Battle of the Pyramids

[22] Joseph P. Farrell, *The Giza Death Star*, pp. 167-168.

With the French victory at the Pyramids, Napoleon quickly established himself as master of Egypt, and the French scholars spread out over Egypt and Giza, compiling data. The well-known Rosetta Stone was discovered by a captain Hautpoul and later deciphered by the French linguistics scholar Champollion, allowing the ancient Egyptian hieroglyphic script to be accurate translated for the first time in almost two millennia.[23]

Napoleon himself, it is alleged, wasted no time in visiting the interior of the Great Pyramid himself, and according to some versions of the story, stayed alone in the King's Chamber, whence he emerged ashen and trembling. At the end of his life, while exiled on St. Helena Island in the South Atlantic, Napoleon is alleged to have said that he saw the entirety of his life, and its end, while in the chamber.

In any case, it *is* clear that Napoleon did visit the structure, and that his soldiers shot their pistols inside the Grand Gallery, and discovered that the structure echoed and re-echoed with the sound of the pistol reports, demonstrating the Gallery's well-designed acoustical properties.

A Contemporary French Drawing of Napoleon and his Officers inside the King's Chamber with the Coffer clearly visible.

---

[23] Captain Hautpoul was from the same family that plays such a prominent part in the alleged mysteries surrounding the French Languedoc town of Rennes-le-Chateau and its local village priest, Berengar Saunière.

French Etching of Napoleon at Giza

For our purposes, however, the French expedition made another discovery, one so obvious that it is often overlooked. In the many etchings the French published in the wake of Napoleon's expedition, one finds an etching of the Great Pyramid, with the Sphinx in the foreground. The etching is unusual in that it appears to be an attempt to depict the Great Pyramid and the Sphinx as they might have looked when were originally constructed, rather than in the degraded condition Napoleon's soldiers and scholars found them in. We know this because the French discovered that the Sphinx was nearly completely buried in sand, the head and face alone protruding from the dunes. The Pyramid itself, of course, had long been stripped of its casing stones which were used for the construction of mosques and other buildings in Cairo.

Yet, in the etching, the Sphinx is depicted not in a desert, but at the edge of a hilly but apparently temperate country, with the Pyramid in the background *and clearly showing a line from base to apex along the apothem*, two points which clearly imply that the French scholars had concluded that the climate was one much different, and that they had noticed the "parabolic" faces of the Pyramid.

French Etching of The Sphinx and the Great Pyramid, with the Pyramid Showing the clear line along the apothem.

If one removes the Sphinx from the picture, the image and feeling conveyed is one of a very modern, and even hardened, military site.

Indeed, as I pointed out in numerous interviews since the publication of the original *Giza Death Star* trilogy, the resemblance of the Great Pyramid in the background of the Napoleonic French etching to a modern *military* installation is eerily anachronistic, as the following pictures of the Anti-Ballistic

Missile phased array radar installation near Nekoma, North Dakota, make abundantly clear. And these pictures also prompt inevitable questions: is there a *physics* reason for the pyramidal shape of such radars? Was this reason derived in some way from a secret study of the Great Pyramid? And is the palpable resemblance of the Napoleonic etching to the Safeguard Anti-Ballistic Missile Phased Array Radar shown below more than just an eerily anachronistic coincidence? Could it rather be in a rooted in parallel science? And is the eerily palpable resemblance of the Giza Compound with its "causeways" and "temples" to the "causeways" and "temples" of the North Dakota installation more than just mere coincidence? Is it simply coincidence that the current faces of the Great Pyramid are "parabolic",[24] and that the faces of the Phased Array Radar pyramid are aligned to four directions of the compass? As we shall argue in the remainder of this book, none of this is a coincidence.

The "Safeguard" Anti-Ballistic Missile Phased Array Radar at Nekoma, North Dakota.

---

[24] It should be noted that when the remaining casing stones of the Great Pyramid were uncovered in the later 19th century, it became clear that the casing stones did *not* reproduce this indentation or "parabolic" feature, and it would thus not have been visible in the Pyramid's original and pristine condition.

Another Ground View of the defunct "Safeguard" Anti-Ballistic Missile Phased Array Radar at Nekoma, North Dakota, complete with "Obelisks" on the Left

An Aerial View of the "Safeguard" Anti-Ballistic Missile Phased Array Radar Site in Nekoma, North Dakota, with its "Pyramidal" reception disk at the top, and the various "causeways" and "temples" of the installation

D. The Thesis

So what, *exactly*, does the weapon hypothesis say? What is the "thesis statement" or "abstract" or "preçis"? In the preface to the original *Giza Death Star* I asked precisely that question in almost the exact same way: "So what does the technical jargon of the weapon hypothesis have to say?" And I answered it this way:

> It says that the Great Pyramid was a phase conjugate mirror and howitzer, utilizing Bohm's "pilot wave" as a superluminal carrier wave to accelerate cohered electromagnetic and gravito-acoustic waves to a target via harmonic interferometry. That rather tangled idea leads to a set of putative principles of its engineering. Since so many of its dimensional measures appear to be harmonically resonant to each other, the Pyramid, as a coupled harmonic oscillator, seems constructed of several oscillators nested within the structure in such a fashion as to suggest a set of feedback loops being used to amplify that oscillated energy.[25]

Were I to rewrite that today in such a way as to convey the same ideas in what amounts to no less a complicated tangle, I would say that *the Great Pyramid is a receiver, transformer, and broadcaster of longitudinal waves—or if one prefer, gravito-electric or electro-acoustic waves— in the medium of local space-time, that is to say, it is a coupled harmonic oscillator of such waves. In order to receive, transform, cohere and broadcast such longitudinal waves, that is to say, in order to function as a coupled harmonic oscillator of the medium itself, the building must reproduce as many dimensional analogues of the local structure of space-time within the structure as it can in order to function as a phase-conjugate mirror of them. As such, its principal means of "targeting" is reliant upon non-local entanglement via harmonic resonance (another reason for all the dimensional analogues).*

This, of course, is only marginally clearer than the original, but amounts to the same thing. It highlights the difficulty of having to reconstruct a weapons function from the structure and from the relevant texts, and to reconstruct a physics upon which it may have

---

[25] Ibid., p. ii.

been based from various obscure areas in contemporary theoretical and applied physics, areas themselves that in some cases skirt the edges of "physics orthodoxy."

In any case, it will be evident to the reader that any explanation of the Great Pyramid along "scientific" lines runs into a problem: how do we know the science posited for the structure is, indeed, the scientific knowledge used in fact? Could the principles behind the structure, along whatever hypothesis that is adopted to explain it, not far exceed our own scientific knowledge?

The answer is of course a firm "yes," and this will become particularly evident as we seek to flesh out the weapon hypothesis by pressing various little known aspects of science literally to the snapping point, and well beyond present capabilities embodied in those aspects and principles.

Indeed, in the second book of the original trilogy, *The Giza Death Star Deployed*, I devoted a whole chapter (chapter five) to the implications of precisely this problem viewed in relation to the evidences adduced for the various hypotheses of the Great Pyramid's intended function. Over the next few pages, I reproduce that chapter exactly because of its intrinsic importance to the argument. This lengthy citation begins at the bracket "[" and ends at the bracket "]", and includes all original block quotations, numbered points, and so on, exactly as they appeared there. The footnotes remain the same in terms of their *content*, but obviously have different *numeration* here in *this* book. Where I have added editorial comments or explanations, these are distinguished by **boldface** parenthetical comments in the main text.

[1. Message? Or Machine? A Preçis of the Weapon Hypothesis (Chapter Five of the Original Giza Death Star Deployed)
a. Five Hypotheses, Two Models, and a Method

(In the opening pages of this chapter) I have referred to several of the most prevalent alternative hypothesis of the origins and functions of the Great Pyramid. For convenience, one may narrow these down to five basic hypotheses:

1. The Time Capsule Hypothesis **(which overlaps with the "bureau of standards" in stone, as a depository of ancient knowledge);**
2. The Prophecy-in-Stone Hypothesis, or as Robert Bauval puts is, a "hermetic device" **(overlapping with the idea that the Pyramid is a psychotronic device directly manipulating human consciousness);**
3. The Observatory Hypothesis **(which overlaps with the Prophecy in Stone hypothesis);**
4. The Machine, or Power Plant Hypothesis;
5. The Weapon Hypothesis.

As alternative explanations of the possible function of the Pyramid, each of these has its unique strengths and weaknesses. And they all share one common strength: none of them adhere to the ridiculous notion that the three giant pyramids of Giza, and in particular, the over-engineered complexity and work of genius that is the Great Pyramid, were **(intended to be)** tombs for dead Egyptian kings. **(Of the three large pyramids, the smallest, Menkaure, is the only one ever to have a corpse clearly associated with it).**

A closer look at the five basic explanations will disclose that Hypothesis 2 is but a variant of 1, and 5 but a variant of 4. Those two groups, 1 and 2, and 4 and 5, view the Great Pyramid by one or another of two models: it is a Message, or it is a Machine. Sandwiched somewhere between these two groups is Hypothesis 3, the Observatory Hypothesis, for the Time Capsule Hypothesis, Hypothesis 1, depends in some degree on the fact that at the very least the Pyramid was a simple machine: an astronomical and terrestrial observatory, a kind of very bulky (and expensive!) sundial.

(1) Weaknesses of the Message Model

In its most notorious form, the "Prophecy-in-Stone" Hypothesis, the Pyramid is viewed as a message… from God, or the gods, or even our more enlightened and sophisticated ancestors, to us. And in most versions, the "message" is understood to be a benign one. We may rather quickly dispense with this view on the basis that those who so view it expose themselves to two serious weaknesses **(and criticisms):**

1. Those who view it as a kind of "Bible Prophecy-in-stone" can do so only by theological interpretation of the biblical text itself, and then comparing that to certain measures of the Pyramid and various historical events. Usually these interpretations run along classic fundamentalist "dispensationalist" lines, complete with the "two-stage" return of Christ, with raptures, tribulations, millennium and various "previous dispensations." Since each of these views are recent innovations, and not part of the original or early Christian patristic understanding of biblical texts, such views are questionable on theological and historical grounds alone. Thus, even if such correspondences *do* exist, and even if this model is "successful" at making predictions of future events, those events can only be successfully "predicted *a posteriori*, that is, they will still have to be "interpreted" as having *been* predicted. *Such correspondences therefore cannot mean* what such interpretations maintain they mean since such "Bible Maps of the Ages" were unknown in the Patristic period and found no entrance into Christianity until Joachim of Fiore in the **(western)** Middle Ages.
2. Those who view it as a type of "New Age Prophecy-in-Stone" are subject to a similar criticism, and even more so, since the selection of various esoteric and mythological traditions is itself an act of constructing the interpretation.

One may therefore dispense with the Prophecy-in-Stone Hypothesis since it depends on supplemental texts or some other supplemental knowledge to convey the message apart from the monument itself. This would introduce "transmission errors" into the message and hence would be self-defeating. Moreover, there is nothing to indicate **(the criteria by)** which corroborative the text or religious tradition (by which the Great Pyramid-as-prophecy is to be interpreted) is to be selected. The monument thus becomes a superfluous, unneeded entity, a six million-ton violation of Ockham's Razor! **(We shall see, however, some texts *do* contain clear references to the Giza plateau, but not as a "prophecy-in-stone," but rather, as an historical record and inventory.)**

A much more sophisticated version of the Message Model is the Time Capsule Hypothesis. On this view, the Pyramid literally becomes, at the minimum, a stone "bureau of standards and measures," and at the maximum, a mathematical message left for us to decipher when we had once again attainted a similar plateau of scientific achievement. One may appreciate the sophistication of this view by comparing it to proposals advanced for contacting and communicating with an extraterrestrial intelligent life form, or conversely, how "they" might attempt to contact and communicate with us.

Obviously, we cannot send a message in English, Arabic, or Japanese, for while such a message *may* be understood as intelligent communication by "them," unless it were an extremely long message, there would be no guarantee that it had been deciphered and understood correctly. The same would be true in reverse. If "they" were communicating to us in ET-German or ET-Swahili, the message would have to be long enough to permit some degree of decryption. And again, there would be no guarantee that the message was correctly translated and understood. Without a corresponding "key-text", we would be virtually lost. We would, in fact, find ourselves back to the problem inherent in the Prophecy-in-Stone Hypothesis: the Monument (in this case, the message itself) would require a text (in this case, the key code) to decipher.

The way around this conundrum is that any such "message" would have to occur in the most universal language possible: mathematics and geometry **(and therefore, very possibly, *music*)**. This is essentially the argument of the two foremost schools of the Message Model of ancient monuments: Richard Hoagland and his Mars researchers on the one hand, and Graham Hancock, Robert Bauval, and the "ancient catastrophists" on the other.

In the Hoagland-Mars version, the "message" is basically that the mathematics and geometries of Cydonia (and by extension, Giza and other such sites on Earth) encode a hyper-dimensional "tetrahedral" physics, a physics that in turn is the key to unleash enormous amounts of energy, if we but knew how to engineer it.[26] While this message is essentially benign, more recently Hoagland's Mars version of the model has taken on elements of the planetary catastrophism of Van Flandern and the Bauval-Hancock version of Giza.

In the latter version, the planetary catastrophe comes in regular, measurable, predictable "cycles" **(that are)** based on astronomical data. On this view, the Giza compound was built as a permanent monument and message by a dying civilization to warn us of a similar fate so that, with the proper enlightenment, preparation, and "spiritual insight", we might avoid or at least mitigate it.

The Hoagland-Mars and the Bauval-Hancock versions of the message model have common features. Both argue for the existence of a paleoancient[27] (and in Hoagland's version, spacefaring) Very High Civilization. Both argue for a major planetary catastrophe on Earth, or Mars, or elsewhere, and in some versions, all three. But for all their resemblances, it is actually Hoagland's version that is the stronger of the two, for the following reasons.

In the Bauval-Hancock version, the problem of textual construction again intrudes, albeit very subtly, in the form of comparisons between Central American monuments, belief systems (the extra-monumental "text" again), and Giza. In Hoagland's version, there really is no extra-monumental text that is subject to construction by which the monuments are interpreted.

But again and again, in his writing and public appearances, Hoagland refers to the "message" of Cydonia as an attempt to

---

[26] In fairness to Hoagland, the zeal and passion with which he expounds the essentially benign nature of this message is perhaps due to the fact that he cannot but help be aware of the fact that the physics he proposes could equally be the source of a great superweapon. Perhaps, then, his zeal and passion are designed to publicize as widely as possible the implications of "tetrahedral physics" before the subject is quietly and deliberately shuffled to the sidelines, only to disappear and become the subject of more sinister research.

[27] In the original books, I coined the redundant term "paleoancient" to denote a high civilization pre-existing the ancient classical civilizations, i.e., a civilization that would have been "ancient" to *them*.

preserve and communicate his tetrahedral "hyperdimensional" physics. Either Hoagland is being merely rhetorical, and no message was intended for us, or Hoagland—if he *means* his rhetoric—has made an assumption that exceeds the evidence.

In short, the basic problem with the Message Model is simply this: the presence of redundant mathematical and geometric relationships may be an indicator of intelligent design, but it is not thereby simultaneously an indication of attempted communication. Any hypothetical reconstruction of the motivations of the builders can only come as a corollary of an analysis of the possible functions of the structures, and any functional analysis must introduce external factors that exist outside of the formal mathematical theorems embodied in the structure. In short, the Message Hypthesis is rather weak, since it violates, or rather, ignores the implications of Gödel's incompleteness Theorem. To maintain that the Cydonia or Giza structures were *intended as communication* is an *a posteriori* argument and therefore a weak argument.[28] Gödel's theorem in fact exposes a serious weakness in any argument that universal mathematical symbols **(or truths)** can function as a message to or from extra-terrestrials beyond the mere communication of "here we are and we are intelligent because we know this." Any formal system remains incomplete, and points to information outside that system, and hence, the Pyramid, if it is such a "statement" or "message in geometry" must perforce be an incomplete one. It *invites* interpretation or "completion" by propositions lying outside the system itself.

This brings us to the last three hypothesis: 3 Observatory, 4 Power Plant, and 5 Weapon.

(2) The Machine Hypothesis

As mentioned above, in all five hypotheses the Pyramid does function as a simple machine, as an observatory of astronomical and terrestrial data.

(a) The Weakness of the Observatory Hypothesis

---

[28] Notice I said a *weak* argument, not an invalid one. *Any* hypothesis concerning an artifact is perforce to some degree *a posteriori*.

But when considered on its own—i.e., as if it were intended to be nothing *more* than an Observatory—a serious weakness exposes itself. Why engineer so many *other* physical and astronomical analogues into the structure? **(That is, what function do they contribute, beyond redundancy, to the function of an observatory?)** Why build into the structure analogues of the atomic weights of certain elements? **(We certainly do not do this for a standard optical or radio observatory!)** Why make the inside of the Observatory **(for all intents and purposes)** inaccessible? **(After all, the ninth century Caliph Al-Maimoun had to *dynamite* his way into the "Observatory".)**

Thus the Observatory hypothesis—while advanced by some of the most sophisticated minds, and seemingly the most rational and easiest of all the alternative hypotheses to swallow—is actually the *least* rational of them all, for it simply ignores the over-engineered complexity of the structure.[29] This brings us to what are, to my mind, the most serious contenders, Dunn's Power Plant Hypothesis, and my own **(elaboration of Sitchin's)** weapon hypothesis.

(b) The Weakness of the Power Plant Hypothesis

Briefly, the weaknesses of Dunn's Power Plant Hypothesis are four:

---

[29] As Gantenbrink also pointed out, the "air shafts" could not possibly have been intended to "observe" the stars since the bends in the shafts made it impossible to view them. He goes too far, however, when he dismisses the stellar alignments of these shafts on that basis. **(To put it differently, an "observatory" and an "observer" are not the same things. One need not assume that the "air shafts" were intended to be utilized by human observers peering at certain stars on certain dates, which in any case they would not have had easy access to, as the case of Caliph Al-Maimoun itself demonstrates. The crookedness of the shafts does not invalidate their alignment on certain stars at certain times, and in that sense the Pyramid *itself* constitutes the "observer" as well as the "observatory.")**

1. It ignores relevant textual data that indicate a possible function of the machine;
2. It ignores relevant textual data that indicate what the missing components might have been; and,
3. It tends to ignore the possible functional purposes of the various mathematical dimensions of the structure and stellar and galactic correspondences of those dimensions.
4. Finally, in seeking to explain the functional purpose of the structure's various components—the chambers, shafts, coffer and so on—Dunn does not state how the acoustic, microwave, and peizo-electric components were *integrated*, nor does he offer an explanation of how embedded quartz crystals in granite could build a charge of more than a few milliamps, far too low, it would seem, to have functioned as a power plant. However, it is to be noted that, suggestively, he mentions Tesla in connection with his theory, implying that he is *not* thinking of its electrical power output in any conventional sense.

This being said, these also constitute in my mind the great strength of his work, in that he seeks to explain the structure in terms of known engineering principles, and not on extrapolated speculative principles of what I call "paleophysics" or other external factors.

(c) The Weapon Hypothesis

The Weapon Hypothesis, conversely, *does* seek to take into account the direct and corroborative textual and archaeological data. Because of this, the Weapon Hypothesis is subject to same type of criticism as advanced against the hypotheses of the Message Model. However, in the Weapon Hypothesis, the reliance upon external text is severely constrained for the following reasons:

1, The direct textual evidence indicative of a weapon function is used as a point of departure to formulate the hypothesis to be investigated, and nothing more. **(And**

as will be seen, the texts themselves *strongly suggest* that they refers to the two giant Pyramids at Giza, as well as clearly stating a weapon function for the Great Pyramid.) That is, Sitchin's "text" does *not* control the details of the investigation of the various parts of the structure, as it does in the Bauval-Hancock version of the Message Model, **(or as texts in general control the interpretation of structural dimensions in the "Pyramid-as-Prophecy" model).** Once the hypothesis is formally announced, science takes over as a basis on which to speculate on the type of physics **(presented in the texts)** and to posit the hypothesized functional purpose of the components of the structure. The hypothesis does *not* therefore stand or fall on the translation or interpretation of the text itself. Text is a corroborative, not primary, datum on this view.

2. Corroborative texts are thus utilized only insofar as they are tools to verify or deny the possible existence of a sophisticated "paleophysics", a physics sophisticated enough to be weaponized in the manner suggested by the primary component of the investigation, the Pyramid itself.

(d) A Method

This being stated, however, there is a substantial method-logical weakness to the Weapon Hypothesis:

1. The Weapon Hypothesis tends to the view that *every* dimensional measure or structural component **(of the Great Pyramid)** was functionally significant, seeking explanations of these putative functions in contemporary research and theory, and when these fail, a speculative reconstruction is attempted. Such may *not* have been the case, and Dunn's hypothesis avoids this weakness.
2. This is mitigated by the fact, however, that the Weapon Hypothesis is an open-ended hypothesis, i.e., it

acknowledges that not every structural feature may ultimately prove to be functionally significant.

The great merit of *both* hypotheses of the Machine Model is their open-endedness. That is, **(some specifics of)** both hypotheses may ultimately be disproven, but in the process, they both have laid a solid *prima facie* case that the Great Pyramid *is* a machine. If the **(details of the)** hypotheses are **(eventually)** counter-indicated **(by fresh discoveries or data)**, it only remains to discover what kind of machine it actually was.

The great strength of Dunn's Power Plant Hypothesis is its reliance solely on an analysis and interpretation of its engineering, apart from external corroborating textual data, whether direct or indirect. By the same token, that is simultaneously its weakness, since it methodologically **(isolates the structure from such evidence)** and imposes the condition that a class of evidence claiming relevance to the question at hand, namely texts, is ignored.

The great strength of the Weapon Hypothesis is that fact that it seeks textual corroboration, and therein lies also its weakness, for to do so it must **(to some degree)** re-interpret those texts, claiming a decayed pedigree from a purely scientific archetype, with all the correspondent historical reconstruction that this entails.][30]

2. Assumptions

The weapon hypothesis of the function of the Great Pyramid rests on the following assumptions:

1. That Nikola Tesla's impulse magnifying transformer technology, and the phenomena it accessed, were the first observed effects of scalar wave stressing of the vacuum potential, or of cohering the "zero point energy flux" of spacetime, **(that is to say, his technology**

---

[30] Joseph P. Farrell, *The Giza Death Star Deployed* (Kempton, Illinois: Adventures Unlimited Press, 2003), pp. 141-148, boldface parenthetical comments added.

produced "electro-acoustic" longitudinal waves in the medium itself);

2. That this technology was the first applied example of a type of physics now known as "scalar electromagnetics" or "electrogravity";
3. That the theoretical foundations of this physics in turn closely parallels the physics exhibited in certain ancient esoteric texts, among them the *Hermetica* of Hermes Tristmegistus… and Chinese traditions of a super-weapon called a yin-yang mirror";
4. That on the basis of a comparison of the known requirements for Tesla's system of the wireless transmission of power for peaceful purposes with the Great Pyramid itself, the latter exhibits certain design features suggesting it was engineered for no other purpose than that of offensive strategic weaponry of mass destruction, and that it was *not* designed merely as a peaceful "power plant" for the wireless transmission of power;
5. That the observed and/or predicted physical effects of this technology and the underlying theoretical model would result in gravitational anomalies, a prediction corroborated by various ancient texts and traditions; and that the use of recursive "time reversed" phase conjugate waves closely parallels myths of "yin-yang" mirrors and other persistent dualistic physical models in ancient cosmological schemes.

The weapon hypothesis must therefore also give a detailed comparison and explanation of each of the following components of the Great Pyramid and their putative functions based on the Tesla-scalar **(electroacoustic or electrogravitic)** model.[31]

3. Models of Weaponization

In the chapters which follow, various models of and ancient allusions to weapons of mass destruction will be discussed, including the following:

[31] Joseph P. Farrell, *The Giza Death Star*, pp. 149-150.

1. "Plasma cannon";
2. "Fusion toches";
3. Solar mirrors;
4. Weaponized weather modifications;
5. Behavior modification;
6. Electro-acoustic vibration of the earth, inducing earthquakes;
7. "Yin-Yang" mirrors relying on recursive waves, i.e., phase conjugation directing superluminal longitudinal waves in the medium that cohere vacuum flux, waves modulated by gravitational, acoustic and electromagnetic information, which could potentially induce nuclear chain reactions directly in the nuclei of atoms in the selected target region. Targeting of such a type of weapon we speculate would be via a kind of harmonic interterferometry and non-local "harmonic entanglement."[32]

E. [The Hypothesized Functions of the Great Pyramid's Component Materials, Geometric and Harmonic Features and Dimensions, Chambers, and Feedback Loops

Assuming each **(material and dimensional)** component of the structure to be essential to its proper and efficient functioning, the weapon hypothesis will have to give explanations for four basic elements of the structure: (1) the selection of particular materials in its construction, including the speculated missing components; (2) its geometric and harmonic features, such as its "parabolic" faces, squared circle and cubed sphere geometries, **(celestial and**

[32] It is to be noted that this section varies somewhat from the original in *The Giza Death Star Deployed*, in that I have attempted to clarify point number 7 with respect to the "Yin-Yang mirror" via harmonic entanglement and phase conjugation. Needless to say, that while the *concepts* being outlined have appeared in public, the technology to utilize them practically in a weaponized fashion as I am proposing remains speculative and conjectural.

geodetic analogues of its dimensional measures) and so on; (3) its internal chambers, and (4) its "feedback" loops.

In order to prepare the reader, each hypothesized function of the components of the structure explored in the remaining chapters are summarized here....

1. The Selection of Materials Used in its Construction

- The Role of ***Hydrogen Plasma*** in the Great Pyramid was to accomplish one or more of the following:
  - Experiments demonstrate that in bucking electromagnetic fields and caduceus coils hydrogen is a necessary technology to access quantum vacuum or Zero Point Energy fluctuations and to engineer the local spacetime structure of the medium, or aether;[33]
  - It is an analogue of nuclear processes in order for the Pyramid to have functioned as a coupled oscillator to those processes in Any Possible Receiver, (i.e., target).[20]

---

[33] As I pointed out in my book published some years after the original *Giza Death Star*, this very point was at the heart of the Nazi wartime and post-war experiments in fusion by Dr. Ronald Richter. Q.v. Joseph P. Farrell, *The Nazi International* (Kempton, Illinois: Adventures Unlimited Press, 2008), pp. 249-352 See especially Richter's comment to the secret U.S. Air Force delegation sent to interview him (p. 343): "(We) assume, that highly compressed electron gas becomes a detector for energy exchange with what we call zero point energy, zero point energy derives from the exclusion principle, the exclusion principle derives from empirical data, in a shock-wave-superimposed, turbulence-feed-back controlled plasma zone exists a high probability for cell-like super-pressure conditions...on the basis of exchange coupling, it seems to be possible to 'extract' a compression-proportional amount of zero-point energy by means of a magnetic-field-controlled exchange fluctuation between the electron gas and a sort of cell structure in space..., representing what we call zero point energy, it seems even possible that the large-amplitude fluctuation signals derive from a mechanism of energy-conversion unknown to us.... plasma implosion analysis might turn out to become an approach to a completely new source of energy." It should be noted also that Richter believed such plasmas under stress were sources of infrasound waves. Q.v. *The Nazi International,* pp. 269-270, 293-294, 316.

- The Role of ***Materials Selected in Construction:***
  - ***Granite*** was selected for:
    - Its known piezoelectric properties in order to build up an enormous electrostatic potential to be accessed and released in a direct current impulse against a resistance barrier of non-linear material **(i.e., the stone itself)** to create a cohered and pulsed] longitudinal "electro-acoustic" [wave.
    - In this regard experiments with caduceus coils wound around quartz crystal and utilizing bucking electromagnetc fields to produce gravitational anomalies may be relevant, since the ancient texts speak of gravitational anomalies in the proximity of the Pyramid when it was fully functional.
    - That lattice structure of granite (i.e., of its quartz) may therefore may have served the function of a wave-guide for the impulse, **(and the varied thicknesses of the stone courses from bottom to top of the structure strongly suggest that their variable thicknesses served the function of a wave-guide as well. That is to say)**, the overall crystalline structure of the Pyramid *itself* may also have functioned as a wave-guide for the impulse.
  - ***Limestone (Calcium Carbonate)*** was selected for the bulk of the structure and for its ***Casing Stones*** because:
    - Non-linear materials are essential in the ... wave-mixing necessary to the engineering of] **longitudinal electro-acoustic** [waves.

2. Its Geometric and Harmonic Dimensional Features such as "Parabolic" Faces, Squared Circle and Cubed Sphere Geometries, Solar and Geodetic Analogues, and So On

- The overall functions of the whole structure of the Great Pyramid are **(speculated to be)** as follows:
  - It functions as a phase conjugate mirror **(and as)** an analogue or coupled harmonic oscillator of the Earth and other local inertial systems.... The "squared circle" and "cubed sphere" properties of the structure make it a

crystalline analogue of the Earth and of Any Possible Receiver (i.e., target) on the Earth or in nearby space.

- Its "parabolic faces" function to collect and focus the background radiaton... of the base galactic and solar systems[34] into the structure and thereby to oscillate and modulate its impulse to Any Possible Receiver.... This design feature confirms its function as an "analogue computer" and "coupled oscillator" of Any Possible Receiver.
- The apothems of the faces **(possibly)** function as the virtual leads to the impulsing transformer component of the structure.
- The missing crystals in the interior of the structure, **(i.e., in)** the Grand Gallery **(and possibly in the Queen's Chamber, and possibly the Coffer of the King's Chamber)**], what I will call ϕ or "phi" crystals [a ϕ-corundum (i.e. sapphire) crystal—and of the apex, function as the electro-gravitational] resonators [of the structure. **(It should be stressed that these putative missing components, with their putative and highly conjectural properties, constitute profoundly speculative hypothetical technologies.)**
- The overall geometry of the Pyramid **(as noted before),** is itself that of a large crystal lattice, **(and specifically that of the tip of a quartz crystal)**, and also possible **(the analogue of a)** coil, and a waveguide, for the impulse.

3. The Functions of Its Internal Chambers

- The Role of ***Construction Features:***
  - The ***Stone Courses***—in their variable heights and thicknesses—**(possibly)** function as analogues and therefore as coupled harmonic oscillators of the atomic weights of **(some of)** the elements as they might occur in Any Possible Receiver.
  - The ***Stone Courses*** also **(possibly)** function as the "windings" of the secondary in the system of Tesla direct

---

[34]By "base galactic and solar systems" I mean "fundamental" in the harmonic sense, i.e., the "base galactic" system is obviously the Milky Way, and the "base solar system" is our own Sol and its planets, comets, asteroids, &c.

impulse magnifying transformers, with the non-linear nature of the limestone itself contributing to the over-all phase conjugation function of the structure. Viewed both as a crystal *and* a coil, the Pyramid's stone courses thus function as a wave-guide for the impulse, with the thicknesses of each course perhaps being analogues of the atomic weights of certain elements.

- The ***Queen's Chamber*** functions, as **(we shall see in our examination of Christ Dunn's Power Plant Hypothesis)** to generate the hydrogen gas. **(By dint of the piezoelectric and impulse transformer functions of the structure),** I further hypothesize that this gas was electrically "pinched" into the ion-acoustic resonant mode of plasma in the other chambers of the structure.
- The ***Grand Gallery*** functioned for electro-acoustic **(or if one prefer electro-gravitic)** infrasound generation, **(modul-ation),** and amplification and to oscillate and pinch the hydrogen gas into the ion-acoustic mode of plasma.
- The ***Antechamber*** functioned as a sonic baffle to "damp" frequencies not resonant to A Given Receiver (i.e., a specific target) allowing only resonant harmonics of the target's "fundamental" electro-gravitational] signature [to enter theKing's Chamber for phase conjugation, cohering, amplification, and pulsing.
- The ***King's Chamber*** combined several functions:
  - As **(possibly)** the "Tertiary Coil" to the magnifying and impulsing transformer;
  - To pinch the plasma further;
  - To stress the piezoelectric properties of its granite in resonance to the harmonics entering from the Grand Gallery via the Antechamber;
  - To amplify **(and cohere)** the gravito-acoustic signal entering from the Grand Gallery;
  - To cohere and modulate the … impulse in the optical cavity of the Coffer by … wave-mixing **(therein).**

- The ***Nested Feedback Loop Structure of the Queen's Chamber, Grand Gallery, Antechamber, King's Chamber (and structure as a whole)….***

- The “uneven” or **(“rough” and)** “unfinished” look to the stone work of the Subterranean Chamber and the large granite blocks of the King’s Chamber was in fact intentional, being the result of “tuning” the structure to the proper resonances.
  - **(As a resonator of the Earth itself, the “fundamental” of the Pyramid and the King’s Chamber is properly understood to be the Schumann Cavity resonance.)]**[35]

However, before we can examine the texts and the structure, we must first dispense with a monstrous hoax, upon which much of the foundation of Egyptology has been constructed.

---

[35] This entire section appeared more or less verbatim in the original *Giza Death Star Deployed,* chapter 5. Certain passages near the end have been omitted.

2
The Vice of Colonel Vyse: Dispensing with an Egyptological Hoax, and the Problem of Dating the Great Pyramid

"One cannot help but wonder whether Howard Vyse was being utilized by the British secret services and, if so, what interest such authorities might have had in seeing the Great Pyramid attributed to the Egyptian king Khufu."
Alan Alford, *The Phoenix Solution*[1]

A. High Strangeness Surrounds the Great Pyramid[2]

[HIGH STRANGENESS ALWAYS SEEMS TO SURROUND the Great Pyramid. Its wondrous mathematical and physical properties are well known. Its builder(s) are not. There is no equivalent plaque saying "Body by Fisher", no "Made in the Fourth Dynasty" or "Product of Mars" label is attached to the structure anywhere. No carving on a cornerstone proclaims it to be "In Loving Memory of Thoth."

In the absence of any direct testimony as to who its builders were, claimants of every type have been argued for, from Egyptian kings to an endless succession of proponents of prophecies in stone, of stargates or other exotic machines. And I include myself in the latter category. In short, *no* one *knows* who built it, nor

[1] Alan Alford, *The Phoenix Solution: Secrets of a Lost Civilisation* (London: Hodder and Stoughton, New English Library, 1998), p. 117.

[2] Much of the material in this chapter simply repeats verbatim portions of *The Giza Death Star Deployed*, although considerable new material is added. Again, I will resort to the convention of placing material from the previous trilogy of books in brackets—[ ]—followed by a footnote to the original citation. Material added to these sections for short clarification or expansion are placed in bold faced parentheses in the main text.

when, nor why. All we have are inferences, conjecture, and some horrifying suggestions from the ancient texts.

The terrible clarity of some of the ancient texts highlights a puzzling fact, and shines a spotlight on an entirely different cast of characters and set of anomalous questions regarding the Pyramid. And writ large over the playbill and plot synopsis is one, looming question: Why, after the publication of Zechariah Sitchin's *The Wars of Gods and Men* has *no* one except this author stepped forward to investigate the *weapon* hypothesis?[3] Surely such a hypothesis, suggested by the ancient texts, is so radical and so pregnant with huge implications for human science, military technology, geopolitics and history, that someone else would have noticed and undertaken an investigation. It is therefore the deafening *silence* that puzzles, especially when considered against the backdrop of the rather noisy effort being made to assert that the Pyramid and the Giza compound contain some "ancient wisdom" of benefit to humanity. Such musings suggest that some sort of deliberate manipulation of opinion might be occurring with the recent Pyramid research.[4]

This means that an investigation of the investigators may uncover some interesting things, perhaps even things that would directly or indirectly corroborate the weapon hypothesis. After all, if anyone ever previously considered the Great Pyramid to be of any military significance, then one would expect to find recurring military and intelligence interest in the structure.

(As we saw in the previous chapter,) almost as soon as this thought is entertained, one recalls the huge military expedition of Napoleon Bonaparte. A large French fleet of frigates and ships-of-the-line transports thousands of French soldiers and the little corporal himself, along with scores of France's finest scientists,

---

[3] Two other authors allude to the weapon hypothesis, as will be seen later.... However, they do not attempt to investigate the hypothesis, they merely mention it.

[4] Of course there is a beneficial aspect to the physics that would have made the Giza Death Star possible, not the least being the ability to draw energy from the quantum vacuum. But the ancient texts do not speak of the Great Pyramid in anything but military terms.

archaeologists, and linguists to Egypt. But why, in a time of lingering instability in revolutionary France and the attendant international tensions in Europe, is this militarily outlandish scheme undertaken by the nineteenth century's otherwise undisputed master of cold calculation in geopolitical and military affairs? In view of the British mastery of the seas, when the ultimate French retreat from Egypt was a foregone conclusion, why was the expedition undertaken at great expense and great military risk in the first place?

The history of the overt modern scientific investigation of the Great Pyramid is therefore connected, at almost every step, with more hidden religious, military, or even esoteric "occult" agendas. To appreciate this fact, one has only to glance at the one "solid" and "conclusive" piece of evidence tying the Pyramid to the Egyptian king Khufu: the discovery by Howard Vyse of hieroglyphic inscriptions in the chambers above the King's Chamber in the nineteenth century, **(chambers which he himself also discovered).** In this case, the middle is the best place to begin the story.][5]

B. Credit where Credit is (Once Again) Due: Zechariah Sitchin Proposes that a Fraud was Perpetrated

In 2017, some fourteen years after I had published *The Giza Death Star Deployed* and first summarized Alan Alford's case against the genuineness of Colonel Howard Vyse's alleged discovery of hieroglyphics in the so-called relieving chambers above the King's Chamber, chambers which he himself also discovered, a remarkable book by Scott Creighton was published. What made the book remarkable was the thoroughness and brilliance both of its research and of its argumentation that Colonel Howard Vyse did indeed forge the hieroglyphics he claimed to have discovered in 1837 in the so-called "relieving chambers" which he also discovered at the same time.

---

[5] Joseph P. Farrelll, *The Giza Death Star Deployed: The Physics and Engineering of the Great Pyramid* (Kempton, Illinois: Adventures Unlimited Press, 2003, ISBN 978-1-931882-19-4), pp. 60-61.

Here, as elsewhere, however, it was Zechariah Sitchin who once again proved to be ahead of the curve with a penetrating insight that was—also in typical Sitchin fashion—not buttressed with good evidence or argument, for it was in fact Sitchin who first proposed that Colonel Vyse's discovery was a vicious hoax. It was because of the weakness of Sitchin's supporting evidence and argumentation that his penetrating insight was forgotten and ignored,[6] until Alan Alford picked up the trail in 1998, updating and recasting Sitchin's insight with a more reasoned case. We will review Alford's case as originally summarized in *The Giza Death Star Deployed*, before attempting to outline the detailed and brilliant argument of Scott Creighton's *The Great Pyramid Hoax*.

Colonel Richard William Howard Vyse (1784-1853)

[6] Scott Creighton, *The Great Pyramid Hoax: The Conspiracy to Conceal the True History of Ancient Egypt* (Rochester, Vermont: Bear and Company, 2017, ISBN 3978-1-59143-789-5), p. 2.

Alternatve Researcher and Scholar
Zecheriah Sitchin (1920-2010)

C. Alan Alford Takes up the Challenge:
Howard Vyse's Forgery and Military Connection

[The remarkable thing, known to most "revisionist Egyptologists," is the almost total lack of mention of the Giza structures, and particularly the Great Pyramid, in Egyptian texts. …(Such) texts as *do* mention the structures do so only with ambiguous phrases. And like all ambiguous phrases, such "mentions" are subject to interpretation; they might *not* be referring to the Great Pyramid and the surrounding structures at all.

How does one explain this curious lack of mention on the part of a society that was meticulous in its record keeping? The lack of records is almost as much of a mystery as the two great Pyramids themselves; indeed, if they were not currently visible, it is questionable if anyone would believe they had ever existed. The ancient Egyptians simply assumed everyone would know about them and thus there was no need to talk about them, or maybe even they, or the people alive at the time of their building, were not completely "in the loop" as to the reason they were being constructed. Finally, if the weapon hypothesis is true, one must also consider the possibility that they were simply "classified" and that public discussion of them may have been forbidden.

Howsoever one interprets this mystifying relative "silence" concerning the Great Pyramid and its large sister Pyramid, the fact remains that the only *unequivocal* mention of **(the pharaoh)** Khufu in connection with the Great Pyramid is the discovery of hieroglyphs in the so-called "relieving chambers" above the main room of the King's Chamber. These glyphs clearly mention Khufu, and their occurrence in the structure itself constitutes strong contextual evidence that the structure was built by that Egyptian king.

At least, *that is the case as far as the standard line of Egyptology goes.*

But there is a problem, and the problem is *who* discovered these glyphs, and *how* they were discovered.] It was British researcher Alan Alford who in 1998 stated [the case and its implications very succinctly:

> Let us now return to the inscription of Khufu's name inside the Great Pyramid which, as I mentioned earlier, potentially offers the key to dating the entire Giza complex. Since this inscription

> was found inside a part of the Pyramid which had previously been sealed, it is by far the strongest evidence that Khufu actually built it. Is this inscription genuine, or is it a fraud? If it is genuine, we will need to ask how Khufu could have produced such a revolutionary structure. If it is a fraud, we will need to seriously consider a pre-dynastic origin for the Great Pyramid, as indicated by the evidence from radiocarbon dating.
>
> The discoverer of the controversial inscription was an Englishman named Richard William Howard Vyse (1784-1853), who came from a well-to-do military family from Buckinghamshire. Howard Vyse had retired from the British Army as a Colonel at the surprisingly young age of forty-one, with twenty years of so of military service behind him. It is believed his family financed his expedition to Egypt between 1835 and 1837, but a hidden agenda behind this benevolence is revealed by the comments of one of Howard Vyse's descendants that the Colonel was rather better at archaeology than soldiering, and, furthermore, was 'rather a trial to his family.'[7]

Vyse arrived in Egypt when he was fifty-one, and during a time when Egypt seemed abuzz with a discovery-a-day, Vyse was "desperate to make a name for himself."[8]

In February `1837, Vyse made his discovery. The first of the "relieving chambers" above the main room of the King's Chamber was named Davidson's chamber, after its discoverer Nathaniel Davidson who found it in 1765. Vyse, discovering a crack in the granite ceiling, was able to determine (according to one story by poking a reed through the crack) that yet another chamber existed above Davidson's chamber. Using gunpowder to help tunnel his way through the relatively softer limestone, Vyse soon found the four remaining small chambers, similar in all respects to Davidson's chamber, with one exception: writing, *the only writing in all the Great Pyramid*, in chambers even more inaccessible than the others. Within the second, fourth, and fifth

---

[7] Alan Alford, *The Phoenix Solution: Secrets of a Lost Civilisation* (London: Hodder and Stoughton, New English Library, 1998, ISBN 978-0340696156), p. 113.

[8] Ibid.

chambers, Vyse claimed to have found three inscriptions of **(the name or "cartouche")** of King Khufu.

Unfortunately, the whole "discovery" occurred in a context casting a long shadow of suspicion over the whole affair. For one thing, Vyse had almost total control over the compound during his expedition. Thus there were none of the scientific controls normally used in archaeology. As Alford puts it, "There was absolutely nothing to prevent Howard Vyse from committing a fraud."[9] The argument that he did commit a fraud is rather strong:

- All the chambers opened by Vyse contained inscriptions, whereas Davidson's did not;
- All the inscriptions were found on walls *except* the eastern walls which **(Vyse)** had "blasted through," a fact which struck Alford, and **(the present)** author, as just too convenient to be true;
- Vyse's expeditionary logs appear to manipulate dates of the discovery of a stone outside the Pyramid bearing Khufu's name **(or cartouche)** to a point after, rather than before, he made his chamber inscription discoveries.
- The original private and unpublished diaries of Vyse have conveniently disappeared, making it difficult to check for possible manipulation of dates.

And finally, one important fact must not be overlooked. The Egyptian authorities have steadily refused to have the ink of the inscriptions carbon-dated. Such testing would conclusively demonstrate whether the Vyse inscriptions were forged.][10]

As will be seen in the next section reviewing Creighton's more recent research and argument, it is on the last two points that Vyse's forgery is revealed, *because of the recent discovery of original private notes of Vyse on his expedition*, notes that were *not* available to Alford in 1998.

---

[9] Alan Alford, *The Phoenix Solution,* p. 116.

[10] Joseph P. Farrell, *The Giza Death Star Deployed*, pp. 62-64.

Nonetheless, even with only Alford's argument in hand at the time of the publication of *The Giza Death Star Deployed*, I could not [help but conclude that the Vyse inscriptions were forgeries. But why would Vyse have endeavored to perpetrate such a monstrous **(and vicious)** fraud on archaeological science and historiography? A clue not only to why, but to who might have ultimately desired such a fraud, lies in Vyse's *subsequent* career success:

> Finally, we must question whether Howard Vyse was really acting as an independent researcher, or whether he was working for the British Government. It is suspicious that, having retired from active service (on half pay) in 1825, Howard Vyse should have received a promotion to full Colonel in the British Army on 10th January 1837, *during his three month absence from Giza.* It is equally suspicious that he was subsequently promoted to the rank of Major General on 9th November 1846, despite his official 'retirement' during the preceding 21 years. One cannot help but wonder whether Howard Vyse was being utilized by the British secret services and, if so, what interest such authorities might have had in seeing the Great Pyramid firmly attributed to the Egyptian king Khufu.[11]

Why indeed, unless of course one recalls, especially during that time, that the British secret service was almost exclusively the preserve of the Masonic and other esoteric fraternities, and that it had indeed been the preserve of such fraternities since its earliest modern associations with Sir Francis Wallsingham and Sir John Dee, from the reigns of Elizabeth I to James I and beyond.

But why would Masons want to attribute such a structure to Khufu? There seem to me to be two possible answers to this question. Either they wished to establish yet another link between ancient Egypt and their own "quasi-Egyptian" Masonic doctrines and traditions, or they wished to misdirect attention from something else they did not wish the general public to know, such as an ancient science or technology that they may have had in their

---

[11] Alan Alford, *The Phoenix Solution*, p. 117, emphasis in the original.

possession, **(or that they may have suspected some Very High Civilization, lost in the mists of pre-history, may have possessed, and that the Great Pyramid was perhaps an artifact of that science and technology).**

In any case, the possible association of Vyse with more hidden agendas of secret societies and militaries is not a new story. Vyse is one of many such individuals, going back to Newton—himself a member of the Royal Society—and Vyse's own contemporary, Napoleon Bonaparte.][12] Nor does it take very long to discover a vast and hidden Masonic influence in Napoleon Bonaparte's expedition as well, for not only were many of its scientists and scholars members of that fraternity, it has long been suspected that Napoleon himself, and the expedition's inspiration, French Foreign Minister Maurice Perigord de Talleyrand, also had strong connections to the Lodge. And in the case of the French expedition and Newton's interest in the Great Pyramid, the interest was at least in part to unlock the secrets of gravity, as we shall see later in this book.

In any case, it is crucial to understand the *significance* of Vyse's discovery, as well as the significance of it being discovered to be a *forgery*, for in the first instance it is akin to finding a plaque reading "Pyramid by Khufu, built proudly in Egypt by Egyptians," and in the second instance it means the structure is of unknown provenance and probably older than Egypt itself, and that the attempt to associate it with Egypt is thus literally a monumental case of cultural appropriation and fraud.

And that is, indeed, the argument and case made by Scott Creighton in his brilliant book *The Great Pyramid Hoax*.

D. Scott Creighton's Ironclad Case for a Hoax and Fraud

At the outset it should be stressed that what follows is a very high overview of a very detailed and complex argument, an argument that required Creighton to lay it out, step by step, in an entire book. What follows cannot therefore replace that book for an

---

[12] Joseph P. Farrell, *The Giza Death Star Deployed*, pp. 64-65.

appreciation of the depth of research and scholarship it entails. What follows, in other words, is the frame, not the whole building.

At the core of the controversy surrounding Colonel Vyse is his allegation not only to have discovered the relieving chambers above Davidson's chamber, but much more importantly, to have discovered the hieroglyphics in those chambers that spelled out the name or cartouche of the Pharaoh Khufu as the overlord of the construction gangs which allegedly built the Pyramid, gangs whose names also appear in some of the hieroglyphs allegedly discovered by Vyse. Until Vyse's discovery of these hieroglyphs in chambers of the Great Pyramid that had been sealed since its original construction,[13] the only things connecting the structure to Khufu were ancient textual attributions of the structure to him in Manetho and Herodotus, or passing references in the Rosetta Stone, which in Vyse's day had only been deciphered a few years earlier by Champollion after its discovery by Napoleon's expedition.[14] As it was, when Vyse made his "discovery," the cartouche of Khufu was not executed in the fine and precise detail that ordinarily accompanies Egyptian sculpted or painted hieroglyphics, but rather were crudely painted affairs, executed with a type of paint called *moghra,* used in Egypt in ancient times, and still being used there at the time of Vyse's expedition.[15] But there's more to the case for fraud than just "crude execution" of a pharaoh's cartouche and proper name, but in order to see what it is, we have to go back to Zechariah Sitchin, and the first allegation of fraud and forgery in connection to Vyse's discovery.

1. Sitchin's Forgery Claim Regarding Vyse: A Misspelling? Or a Smudge?

Basing his argument on the drawings of an artist accompanying Vyse's expedition, Zechariah Sitchin argued that

---

[13] Scott Creighton, *The Great Pyramid Hoax: The Conspiracy to Conceal the True History of Ancient Egypt* (Rochester, Vermont: Bear and Company, 2017, ISBN 978-1-59143-789-5), p. 5.

[14] Ibid., pp. 5-6.

[15] Ibid., pp. 6-7.

one key glyph in Khufu's name cartouche was that of a tiny smudged circle with a dot in the center of the circle, the glyph for the sun god Ra, and not the glyph for the "kh" in "Khufu." In effect, the glyph was akin to a misspelling and such an obvious error that no ancient Egyptian would ever have made it.[16] Following this clue to the British Museum where he requested to examine Colonel Vyse's published diary, Sitchin was shown the diary facsimile parchments of Colonel Vyse's assistant, J.R. Hill, and again, the name of Khufu was clearly misspelled with the circle and dot—the symbol of Ra—substituted for that of "kh." On this basis Sitchin concluded that Vyse had perpetrated a vicious hoax, and that Egyptology had not caught the forgery.

As can often happen, a brilliant insight can be buttressed with faulty evidence or reasoning, as any researcher knows. That was the case with Sitchin: his conclusion was correct, but his argument was not so much faulty, as not deep enough. In this case, Sitchin relied on secondary sources—those of men accompanying Vyse – rather than an *in situ* examination of the hieroglyphs in the "relieving chambers". In spite of his attempts to examine them, Sitchin never received permission from the Egyptian government to do so. If he *had* visited the Great Pyramid and *had* examined the cartouches closely and personally, he would have discovered that the small and somewhat smudged drawing merely misrepresented what was painted on the chamber walls, namely, a small circle with three horizontal lines inside, *which was, indeed, the correct spelling for "Khufu" when compared to the rest of the glyphs in the cartouche.*[17]

As we will see, however, that was not and is not the end of the story, for as it turns out Sitchin was correct about the idea of a forgery, and was even correct in a very twisted way about the misspelling; the problem was that both the forgery and misspelling were much more *subtle* than Sitchin assumed.

2. Egyptology Strikes Back:

---

[16] Scott Creighton, *The Great Pyramid Hoax,* pp. 13-14.
[17] Scotr Creighton, *The Great Pyramid Hoax*, pp. 14-15.

The "Too Tight to Paint" Argument

One of the strongest and most powerful arguments that Egyptology has for the legitimacy of Vyse's painted hieroglyphics discovery is the fact that many of these glyphs appear in spaces that are "too tight to paint." The quarry marks in Vyse's discovered inscriptions, they argue,

> ...can only be observed through small cracks between tightly fitting adjacent blocks. How would it have been possible, the Egyptologists ask, for any forger to get a brush into the tight gaps between these immovable, seventy-ton blocks and paint any meaningful marks onto the faces of those closely fitting blocks? These marks, they insist, had to have been painted onto the blocks *before* they were set into place (i.e., when the block faces were accessible at the quarries), so they conclude that *all* the painted marks, including those in plain sight, *must* bc genuine.[18]

Note that the Khufu cartouches, which form the core of Vyse's "discovery," are themselves in perfectly visible places in the "relieving chambers," and are *not* located in the "too tight to paint" spaces.

The argument itself is a one of the most peculiar versions of the distributed middle fallacy that I've ever encountered, such that only Egyptology could makeit: my dog is brown, your dog is brown, therefore, your dog is my dog. My dog may indeed be brown, and your dog may indeed be brown, and indeed, they may be the same dog, but one cannot logically reach that conclusion on the basis of "brownness" alone. Other factors must impinge in order to reach that conclusion, and those factors may indicate either that our dogs are the same identical dog, or that they are not.

Similarly, "those paint marks are inaccessible", and therefore "Vyse could not have forged them". But "these paint marks" *are* accessible, but Vyse could not have forged them because the *other* paint marks are *inaccessible* and cannot be

[18] Scott Creighton, *The Great Pyramid Hoax*, p. 12.

painted in the tight spaces.[19] The fact that both sets of marks are "painted" indicates nothing whatsoever on the legitimacy or non-legitimacy of the cartouche name of Khufu or whether or not Vyse painted them.

3. Refuting the "Too Tight to Paint" Argument

However, even the "too tight to paint" argument itself may not be true, says Creighton. Noting that the space between these tightly fitting blocks are about one inch wide,[20] he outlines a theory of yet another Pyramid researcher, Dennis Payne, on how the quarry marks could have been painted in such a tight space. Perry's theory requires some string, some glue, some paint, and two stiff boards, both of which may be slid into the space. One then fixes the string to one board in the shape desired by gluing it in place, then saturating it with paint. Then the board with this string facing the surface to be painted is slid into the crack, and another board is slid in behind that, and then pressed against the first board. The painted string will leave its mark after the two boards are carefully retracted. Creighton quips, "So, once again, what was once held by Egyptologists to be an impossible task is shown to be not so impossible after all."[21]

4. Further Problems for the Theory Of Vyse's Discovery's Legitimacy: "Partials", Contour Fitting, and Paint Runnels

Another problem for the forgery theory cited by standard Egyptology is the fact that some of the hieroglyphics "discovered"

---

[19]Mr. Creighton points out that well-known alternative researcher Graham Hancock was taken in by this argument: having initially supported the forgery theory, Hancock later retracted it in the face of the "too-tight-to-paint" argument, and then later retracted his retraction, noting that the Khufu cartouches were in full view, and not in a tight space, and that Vyse could therefore have forged them. See Scott Creighton, *The Great Pyramid Hoax*, pp. 96-97.

[20] Scott Creighton, op. cit., p. 96.

[21] Scott Creighton, *The Great Pyramid Hoax*, p. 98.

by Vyse appear to be partially blocked by the granite stones of the relieving chambers, with only a portion of the whole hieroglyphic currently visible. Ceighton observes that it's very unlikely we'll ever know for certain as the Egyptian authorities are not likely to allow the stone to be chipped away to see for certain.[22]

However, there is another problem for standard Egyptology when considering these so-called "partials". In some cases, the hieroglyphs not only appear to be partially blocked by the stones of the floor of the chamber, but these sometimes occur *right next to* hieroglyphs that *follow the contour of the existing chamber floor, rather than be partially blocked. Moreover, in some cases the hieroglyphs fill the space between floor and ceiling, or between the floor and the top of a block.*[23] Creighton's argument may not be immediately apparent without stating it explicitly: if the hieroglyphs were genuine and placed *prior* to the construction of the Pyramid, then one would come to expect some "partials." If, on the other hand, the hieroglyphs were added at any point *after* its construction, including by Vyse or his team, then one would expect the "contour fitting" that one also sees. What one actually sees is both types.

But is this not a species of the peculiar Egyptological version of the fallacy of the distributed middle we observed previously with respect to these painted hierglyphics, only in reverse? Answer, no, because Creighton is *not* arguing that all the hieroglyphs in the "relieving chambers" are forgeries, *only that the names and cartouches of Khufu* are, and that they alone, on which the association of the Great Pyramid with Khufu is primarily founded, are forgeries, and that those that show "contour fitting" are likely to be hoaxes as well.

However, having shown all this, Creighton introduces another heavy blow to Egyptology with a consideration of the "paint runnels" that appear on the Khufu cartouche "discovered" by Vyse in Campbell's Chamber. Creighton observes that in 2005 a French photographer named Patrick Chapuis "presented one of

---

[22] Ibid., p. 93.

[23] See the diagram, for example, on p. 100 of Creighton's book.

the most stunning high-resolution photographs ever taken of the Khufu cartouche in Campbell's Chamber."[24] The photograph disclosed that not only had the cartouche been painted from the lower right, counter-clockwise, but that two tiny runnels of paint could be seen running "vertically down to the roof block."[25] Additionally, the photograph revealed what were apparent scraping marks on the roof block to the lower left of the cartouche, "as though to remove something from the surface of the roof block," the "something" being perhaps "where unwanted paint had trickled down from the bottom right-right section of the cartouche…"[26]

While this "paint runnel" discovery does *not* constitute the core of Creighton's argument that Vyse's discovery of hieroglyphics in the relieving chambers connecting the Great Pyramid to Khufu is one gigantic hoax, it *is* worth noting what he says about the significance of the paint runnels on that cartouche:

> What these paint runnels seem to indicate, of course, is that this…cartouche, contrary to what Egyptologists state, was actually painted *in situ*; that is, *not* when the block was cut at the quarry, but *after* the block was set in place. One might not immediately understand the implication of such an observation, but the simple fact is that this finding, all by itself, proves beyond reasonable doubt that this cartouche must then be a modern fake.[27]

One might object that this is not necessarily so, and that carbon-dating of the paint itself might reveal its origin in ancient Egyptian times, and that might indeed be true, but thus far the Egyptian government and its Department of Antiquities has steadfastly refused to allow such testing.

That testing, however, is not likely to be comforting for standard Egyptology, because of what Creighton discovered about the difference between Colonel Vyse's *published* writings, and his *hitherto lost private journal…*

---

[24] Scott Creighton, *The Great Pyramid Hoax*, p. 130.
[25] Ibid., p. 131.
[26] Ibid.
[27] Ibid., pp. 131, 133.

5. Vyse's Private versus Public Journals: Vyse's Foreknowledge of the Proper Writing of the Cartouche of Khufu

The very mention of a *private* and *unpublished* journal by Colonel Vyse is, as we shall see, the key that unlocks the mystery of the Vyse hoax, and how he accomplished it. Creighton observes that Alan Alford himself unsuccessfully sought to find the journal, realizing that "it might reveal some pertinent truths about Vyse's time at Giza that had been omitted by Vyse from his published work."[28] Whereas Alford was unsuccessful, however, Creighton—thanks to the internet—was able to track down and locate Vyse's private journal in the Centre for Buckinghamshire Studies in Aylesbury.[29]

One of the first difficulties encountered when comparing Vyse's public to his private notes is that the private notes reveal his foreknowledge of how to write Khufu's name cartouche correctly, while his public writings do not disclose this fact.[30] A question that immediately occurs is whether or not the correct spelling of Khfu's cartouche was publicly (though not necessarily widely) known before Colonel Vyse allegedly "discovered" the name in the hieroglyphs of the relieving chambers he had blasted into. Fortunately, the answer as Creighton points out is "yes", for in 1832 the Italian archaeologist and scholar Rosellini published his *l'Monumenti Dell'Egitto e Della Nubia* where a version of the cartouche is clearly identified.[31]

The problem begins to focus into clearer outlines by looking carefully at Vyse's private and unpublished notes which were supposedly copied from what he found in Campbell's Chamber, for in the journal one clearly sees at least two versions of the cartouche of Khufu, *both of which differ from that found in*

---

[28] Scott Creighton, *The Great Pyramid Hoax*, p. 158.

[29] Scorr Creighton, *The Great Pyramid Hoax*, p. 159.

[30] Scott Creighton, *The Great Pyramid Hoax*, p. 21. See also the discussion of pertinent details on p 63.

[31] Scott Creighton, *The Great Pyramid Hoax*, pp. 23-24.

Campbell's Chamber. To make matters much worse, the version in the chamber has something present that is *missing* from the versions in Vyse's private journal.[32]

Creighton records the moment that he and his wife realized they finally had irrefutable proof that the whole centerpiece of Egyptology's argument that the Great Pyramid was built in the time of and by the Pharaoh Khufu had completely collapsed:

> ...for before us, on these two pages of Vyse's private journal, was compelling evidence that the cartouche of ... Khufu must, in fact, have been forged by him—just as a number of researchers and writers over the years had suspected. To say that we were dumb-struck by what we had uncovered would be an understatement; here we had found in the colonel's own diary, in his own hand, evidence that proved, beyond reasonable doubt, that Vyse had perpetrated the hoax of all history.[33]

To understand why requires a bit of an explanation of what was written on those two pages, and why it points irresistibly to forgery.

The first datum that confronted Creighton and his wife was the entry of May 27, 1837, where Colonel Vyse had hand-drawn the Khufu cartouche. Remembering that ancient Egyptian is read right-to-left, the first element in the hand-drawn cartouche is a small circle. However, inside of the disk are two dots, rather than the three horizontal lines one actually sees in the cartouche in Campbell's Chamber. What is important to note is that the cartouche written in Vyse's private journal *must* be referring to the cartouche in Campbell's Chamber, because absent the distinguishing (and correct) horizontal lines inside the circle, *all other markings in the journal correspond to those in the Chamber.*[34]

A second entry on June 16, 1837 in Colonel Vyse's private journal, unlike the May 27 entry, renders the Khufu cartouche differently from the first, this time with a simple circle which is

---

[32] Scott Creighton, *The Great Pyramid Hoax*, p. 161.
[33] Scott Creighton, *The Great Pyramid Hoax*, pp. 161-162.
[34] Scott Creighton, *The Great Pyramid Hoax* pp. 162-163.

absent any distinguishing marks on its interior, having neither the "dot" of Sitchin's alleged misspelling of the "kh" as "Ra", nor with the two dots of the May 27 entry, nor with the three horizontal lines of the cartouche that actually appears in Campbell's chamber. To the right of this hand-drawn new version of the cartouche is a written comment, again in Vyse's handwriting, that says "in Campbell's Chamber."[35]

We now have three versions of the cartouche: (1) the cartouche of the May 27, 1837 private journal entry, with the initial circle of the cartouche drawn with two dots inside of it, (2) the cartouche of June 16, 1837, with the initial circle of the cartouche drawn without *any* internal marks whatsoever, and with a handwritten comment next to it that reads "in Campbell's Chamber," and finally (3) the cartouche as it exists to this day in Campbell's chamber, with the initial circle drawn with three horizontal lines inside of it.

At this point, we also note that Vyse's June 16 1837 entry *also* includes a *third* drawing of the cartouche, this time with a double circle as the initial glyph, and inside of *that*, three horizontal lines. There are thus *two* versions of the cartouche for the June 16 1837 entry, one of the cartouche with a single circle with no interior marks, and one with a double circle, inside of which are three horizontal lines. Next to the drawing *without* any interior marks in the circle, Vyse wrote "in Campbell's Chamber," and next to the cartouche with the double circle and horizontal lines, he appended yet another *almost* identical comment reading "Cartouche in Campbell's Chamber."[36]

Additonally, the drawing of the cartouche containing the double circle and horizontal lines also contains other marks. A cartouche itself is a group of glyphs *inside an oval*, and in this case, the cartouche drawn with the double circle and horizontal lines has an "x" placed upon the lower center oval of the cartouche itself. To the upper right, but not on the oval of the cartouche, is placed another "x", and to the right of the double circle is a small

---

[35] Scott Creighton, *The Great Pyramid Hoax*, p. 163.
[36] Scott Creighton, *The Great Pyramid Hoax*, p. 165.

mark almost resembling a "check mark" (√) or an upward slash (/), beneath all of which Vyse has added his comment "Cartouche in Campbell's Chamber".[37]

Creighton asks the obvious question, "why would Vyse draw the … Khufu cartouche in his journal, on two separate occasions, differently from what we actually observe today in this chamber?"[38] *The resolution to this dilemma is rather simple, though it was unknown in Vyse's time. Simply put, the cartouche of Khufu had two spellings, both of which were correct,* one with the initial circle completely empty, and the other with a circle inside of which are the horizontal lines.

> However, in 1837 this fact wasn't yet fully understood, and when Perring presented to Vyse (on June 2) information that showed a slightly different spelling of … Khufu (i.e., a cartouche disc that wasn't blank but contained three horizontal lines), this new information created an ambiguity *and caused Vyse to have doubts over which was the correct disc to use; he wouldn't have known then what we know today, that a blank disc is just as acceptable for the spelling of Khufu as a disc containing striations. This doubt is clearly expressed in Vyse's June 16 journal entry, where a series of edits and annotations are clearly observed on this page.*
>
> Vyse would have well understood that if he had sent a facsimile copy of the Khufu cartouche back to London with what turned out to be the incorrect spelling, he could have been very quickly uncovered as a fraudster and his "discoveries" consigned to ignominy. There was much at stake here. If he could just be the first to empirically connect the Great Pyramid to … Khufu, then his name would be immortalized in history. He had to get the Khufu cartouche right. Should the disc be blank, or should it have internal lines…?

In other words, *Vyse's forgery cannot be detected by the presence of a misspelling; there is no misspelling, but Vyse himself could never have known that since the two spellings of Khufu were*

[37] Scott Creighton, *The Great Pyramid Hoax*, p. 166.
[38] Scott Creighton, *The Great Pyramid Hoax*, p. 166.

unknown in his day, it being assumed that there was only one correct spelling!

It is thus the doubt expressed in his journal entry of June 17, 1837, after another spelling of Khufu was pointed out to him, that Vyse unwittingly "betrays the truth of the situation."[39] What all this means is that it is highly likely that initially Vyse and/or his team had painted the cartouche in the chamber *without* the horizontal lines. Then Perring revealed to Vyse another, and more easily accessible spelling *with* the striations on June 2, 1837. Vyse wavered and deliberated, and recorded the wavering in his private journal. Then, at some point, adds the horizontal lines. Interestingly enough, Creighton states that two individuals who had visited Campbell's Chamber in 2013 noted "that the three lines within the disc of the Khufu cartouche have a slightly different color tone from the rest of the cartouche" indicating two different paint mixes were used, one in May for the initial "discovery:, and another in June, of 1837, for its "correction".[40]

Had Colonel Vyse never worried about being "caught" and recorded his doubts in his private journal over which spelling—both of which as it turns out were correct—his hoax would never have been caught. In the end, Sitchin was right: Vyse did commit a forgery, but was not caught because of a misspelling as per Sitchin, but because of his *fears* about doing so. And in the end, Alford was right, because it was Vyse's private journal which exposed the fraud, but for a reason neither Sitchin nor Alford ever expected.[41]

---

[39] Scott Creighton, *The Great Pyramid Hoax*, p. 175.

[40] Scott Creighton, *The Great Pyramid Hoax*, p. 181. Alford also notes the discrepancy with dates without even knowing of the private diary: "...Howard Vyse's diaries contained a deliberate manipulation of dates which gave the impression that he had found a fragment of stone bearing Khufu's name *after* rather than *before* his discovery of Khufu's name inside the Pyramid." (Alford, *The Phoenix Solution*, p. 116).

[41] Alan Alford also mentions the spelling problem and Egyptology's conclusion that several spellings of Khufu were possible. See Alford, *The Phoenix Solution: Secrets of a Lost Civilisation*, p. 118.

The significance of Creighton's breathtaking research and discovery cannot be overestimated, and it is best to allow him to summarize its significance in his own words:

> … at a stroke, Egyptology loses the key piece of evidence that allows it to attribute the structure to Khufu and to a construction date of circa 2550 BCE. With this vital evidence removed, then the Giza pyramids become monuments whose provenance is much less certain and, as such, reopens the question as to who really was the builder of these monuments, when were they built, and for what purpose.[42]

As we shall see in the next chapter, those questions are seriously answered by Zechariah Sitchin's examination of other ancient Sumerian texts, and by a curiously *un-Sumerian-looking* picture.

Was Colonel Vyse capable of committing such a fraud? An examination of his career would have to indicate a positive answer to that question, for the evidence is clear that he was willing to commit voting fraud to secure his seat in Parliament and then deny ever having done so, long before his expedition to Egypt.[43]

E. Implications: Dating the Great Pyramid, and an Answer to Alford's Epigraph

The epigraph of Alan Alford, which opened this chapter, posed the question of whether or not there was a hidden agenda at work behind the expedition of Colonel Vyse, a hidden agenda that had a military purpose in attributing the Great Pyramid to the time and construction of Pharaoh Khufu:

> .. (We) must question whether Howard Vyse was really acting as an independent researcher, or whether he was working for the British Government. It is suspicious that, having retired from active service (on half pay) in 1825, Howard Vyse should have received a promotion to full Colonel in the British Army on 10th January 1837, *during his three-month absence from Giza*. It is

[42] Scott Creighton, *The Great Pyramid Hoax*, p. 3.
[43] Scott Creighton, *The Great Pyramid Hoax*, p. 37.

> equally suspicious that he was subsequently promoted to the rank of Major General on 9th November 1846, despite his official 'retirement' during the preceding 21 years. One cannot help but wonder whether Howard Vyse was being utilized by the British secret services and, if so, what interest such authorities might have had in seeing the Great Pyramid firmly attributed to the Egyptian king Khufu.[44]

The answer seems rather obvious from the standpoint of the weapon hypothesis itself, and may have been the conclusion someone, somewhere within the military bowels of the British Empire concluded after reading certain ancient texts; best to deflect attention from a possible military purpose by deflecting attention to Egypt.

For an outline of *that* case, however, we must next turn to a consideration of Zechariah Sitchin, and the text which first suggested to him that the Great Pyramid of Giza was at the center of a great "pyramid war" for its control, and that it was "the great weapon." We must look at a Sumerian text with a curiously *Egyptian* looking "visual aid"...

[44] Alan Alford, *The Phoenix Solution*, p. 117.

Ninurta (right), wielding his "thunderbolts" against Abzu (left), who has stolen the Tablets of Destinies; relief from the Assyrian temple Eshumasha

Ninurta's Victory Seal

3
The Paleography of Paleophysics: Ancient Texts, Ancient Wars, and the Origin of the Weapon Hypothesis

"The main merit of this language has been its built-in ambiguity. Myth can be used as a vehicle for handing down solid knowledge independently from the degree of insight of the people who do the actual telling of stories, fables, etc. In ancient times, moreover, it allowed the members of the archaic 'brain trust' to 'talk shop' unaffected by the presence of laymen: the danger of giving something away was practically nil."
Giorgio de Santillana and Hertha von Dechend[1]

"An Akkadian "Book of Job" titled ***ludlul Bel Nimeqi ("I Praise the Lord of Deepness")*** *refers to the "irresistible demon that has exited from the Ekur" in a land "across the horizon, in the Lower World [Africa]."*
Zechariah Sitchin[2]

A. A Personal Note

THE GREAT PYRAMID WEAPON HYPOTHESIS IS NOT, INDEED, MY OWN, but someone else's. It is, in fact, the hypothesis of the late Zechariah Sitchin, elaborated in the context of his book about ancient cosmic wars of the gods and men in a book of that very same name: *The Wars of Gods and Men*.

I first heard about and read this book right around the very time that British-American engineer Chris Dunn published a remarkable book, *The Giza Power Plant*, and began to interview

[1] Giorgio de Santillana and Hertha von dechend, *Hamlet's Mill: An Essay on Myth and the Frame of Time* (Boston, 1969: Gambit Incorporated, no ISBN), p. 312.

[2] Zechariah Sitchin, *The Wars of Gods and Men* (Santa Fe, New Mexico: Bear and Company Publishing, 1985, ISBN 0-939680-90-4), p. 140. It should be noted that the world *ekur* can mean both mountain, or ziggurat, or pyramid.

on many radio shows, including the then famous overnight American radio talk show, Coast to Coast AM with Art Bell. Mr. Dunn's hypothesis was relatively simple: if one examined the structure of the Great Pyramid with an engineer's eye and not the jaundiced and prejudiced eye of "Egyptology" and its narrative, it became clear that the structure was not a funeral structure or elaborate mortuary as held by the popular imagination; it was, rather, a *machine*. What *type* of machine is evident from the title of Mr. Dunn's book, though as we shall see, there are details of his argument that, oddly, square quite neatly with details mentioned by Sitchin. Even more oddly, at the time of the publication of Sitchin's *Wars of Gods and Men*, Mr. Dunn's book had *not* been published, and Mr. Dunn thus wrote his book based solely on his personal examination of the structure, and without any reference to or knowledge of Sitchin's book.

This makes any parallel insights the two might have had all the more remarkable and important.

Most remarkable of all is that even though Sitchin elaborated his weapon hypothesis, no one seemed to have noticed its cosmic significance and importance, including Sitchin himself, who was content to argue that the Pyramid *was* a weapon, that wars were fought over it (and *with* it), but who left unanswered the all-important question of how it might have worked and what type of weapon it may have been.

The purpose of the three earlier *Giza Death Star* books, and of this book, was and is to elaborate (1) *how such a weapon might have worked* and (2) *the type of capabilities it may have possessed.* Obviously, these two topics are interrelated, and in elaborating them a careful course must be steered between too much speculative reconstruction of ancient texts on the one hand, and the typical "academic" attitude that such texts could not possibly contain any accurate scientific or technological information on the other. Typically, such texts are rather viewed by academia as primitive Vedic, Mesopotamian, Graeco-Egyptian or even Meso-American science fiction full of "primitive religious metaphor." If one were to draw an analogy of the approach I assumed in elaboration and construction of Sitchin's weapon hypothesis, I had to put myself in the position of an analyst hired by a government (or a very wealthy patron) to review Sitchin's work and comment on any potentialities that the weapon hypothesis might be true, why it might be true, and most importantly, how it might have worked and what it actually did. My approach remains the same in this book. As such, in this book I examine *additional* texts that I regard as either strongly alluding to the Great Pyramid, or texts that in more general fashion refer to the ancient "cosmic wars" and the horrendous weapons and science with which they were fought. In other words, this book is a kind of "briefing document" or "report."

With that on the record, it is now time to revisit the first of those texts that we encountered in *The Giza Death Star*, and to reprise what I wrote there about Sitchin's reconstruction and interpretation of that text. In doing so, I will cite the major sections of *The Giza Death Star* in which these passages, and my interpretation of them, occurs. Additionally, in citing these passages, I will also cite the original footnote and any annotations therein, retaining our bracket notation "[" and "]" for the beginning and end of those sections..

B. Zechariah Sitchen's Pyramid Wars and the Weapon Hypothesis

> [Sitchin presents credible textual evidence that the Great Pyramid of Giza was the primary component of some form of "paleoancient", though certainly not primitive, weapons system of mass destruction in his work *The Wars of Gods and Men.*[3] This section is a preçis of that textual evidence and a brief analysis of the type of weapon that his analysis of the texts seems to imply.
>
> Sitchin's texts relate to the final events of a paleo-global war,[4] apparently fought, according to him, with nuclear

---

[3] Zechariah Sitchin, *The Wars of Gods and Men*, Book III of *The Earth Chronicles* (Avon Books: 1985), pp. 163-174.

[4] This is the original reading of *The Giza Death Star*, though subsequent books made clear that I regarded this war as not merely global in extent, but potentially, interplatenary in extent as well, as the subsequent examination of the Babylonian epic *Enuma Elish* in *The Giza Death Star Destroyed* made clear. That examination will be reviewed later in this book, along with other textual evidence. One may thus also

and other more terrifying weapons of mass destruction, a war similar in this respect to … wars recounted in the ancient Hindu epics, the *Ramayana* and the *Mahabharatra*. Sitchin calls this war "The Second Pyramid War." The focal point of this war is for control of the Great Pyramid, which is the ultimate weapon. This in itself suggests that its destructive power was far in excess of that imaginable with nuclear weapons, since such weapons were, according to Sitchin, deployed in that war and *used in a subsequent war* after the Pyramid's destruction.

The outlines of that struggle go something like this. The gods of Mesopotamia, eventually victorious over Marduk, who was besieged within the Pyramid itself, dispatch a team to enter the structure, inventory its contents, and designate which components should be destroyed and which should be waved for utilization in other devices.[5]

Thus, if the Pyramid was a weapon, then (adding to our list of …tests):

(1) it was, by definition, some sort of *machine*, a fact which can be verified by careful consideration of its (remaining) components and construction; and,

(2) *other* texts not examined by Sitchin should indicate a similar purpose of function of the Giza complex

---

say that this war was "paleo-cosmic" or at the minimum, "paleo-interplanetary."

[5] That inventory of items slated for destruction and those slated to be used in something else, including a small category of items that could not *be* destroyed, occurs in a Mesopotamian "epic", *The Epic of Ninurta*, portions of which Sitchen produces in *The Wars of Gods and Men*. When I originally read the epic, it became clear that nothing about the work could qualify it as an "epic" in any literary sense, for it is about as exciting as reading the index to the Sears catalogue or the *Encyclopedia Britannica*. I reproduced the text in its entirety in my later book *The Cosmic War: Interplanetary Warfare, Modern Physics, and Ancient Texts* (Kempton, Illinois: Adventures Unlimited Press, 2007), pp. 204-232. The text cited through Sitchin's treatment as reviewed in *The Giza Death Star* and here, however, is the *Lugal-e Ud Melam-bi*, edited and compiled in *Altorientalische Texte und Untersuchungen* by Samuel Geller (q.v. Sitchin, *The Wars of Gods and Men*, pp. 158-159).

or associate such structures with wars and weaponry;[6] and,

(3) some of its components are missing from the structure, a fact which should be in evidence by a careful consideration of the contemporary shell now remaining at Giza.

This list constitutes the first series of Sitchin's "pyramid hypotheses," and it must be carefully considered before proceeding.

First, the texts imply that the structure was conceived and built as a machine, in this case, a weapon of extraordinary power. As such, the verification of this hypothesis will lie in part in an examination of the *form* and *of the materials used* in its construction. Given the number of mathematical and harmonic relationships *alone* that are embedded within it, I postulate that the simplest structure that *could* be built embodying all these relationships in precisely the manner that they were embodied was precisely in the form of a pyramid *and no other*.[7]

Second, Sitchin's texts indicate that the structure was subsequently entered after its completion for the purpose of an inventory and destruction of some of its contents and the removal of other components. The verification of this

---

[6] The bold faced phrase has been added to the text in this book.

[7] The original footnote reads: "That I cannot, obviously, test this postulate goes without saying, for it would require the modeling and computational power of a very large mainframe computer and considerable care and expertise in establishing the parameters of the program." As will be seen subsequently, the pyramidal *form*, as studied by the Soviets during their secret studies of pyramids during the height of the Cold War, revealed certain inherent abilities to manipulate forces by dint of their very *form* or *shape*, and it will be shown that the exact dimensions of the Great Pyramid permit it to perform a *variety* of functions proper both to "flattened" *and* to "narrow and tall" pyramids. Briefly put, "flattened" pyramids act in a similar fashion to satellite dishes, collecting signals and energy. Tall and narrow pyramids act more like broadcast antennae. The Great Pyramid, exactly between the two, can perform *both* functions, i.e., it can collect, amplify, and (re)broadcast energy.

hypothesis will lie in any trace evidence that the pyramid was entered prior to its "modern" forced entry by the Moslem caliph in the ninth century. As will subsequently be seen, Sitchin's texts moreover imply the manner of this entry, and thereby the manner of its verification.

Third, the very suggestion of missing components means the modern structure is but a shell of its true, former, self. This indicates that *some of its functions cannot properly be understood without an exact knowledge of its missing components and their functions.* In the absence of such knowledge, such components as may have been deposited initially within the structure must be speculatively reconstructed on the basis of the existing structure and the physical functions of its various components and nested mathematical, harmonic relationships. Sitchin's texts do provide an important clue as to what those missing components may have been. His texts indicate that the Great Pyramid's primary function was that of a weapons platform. Moreover, they indicate that it was a weapon of such destructive power that it exceeded the power of nuclear weapons. They indicate that its destructive power was so extraordinary that the "victors" in the "Second Pyramid War" ordered its permanent incapacitation; (if this be true) then this means that the shell that remains at Giza is the *secondary* structure. The *primary* components are missing.

Consequently, any theory that attempts to reconstruct its function solely on the basis of "reverse engineering" its remaining shell without consulting those texts, and therefore without due consideration of what those missing components might have been, is an inadequate theory. In this respect, Bauval, Hancock, and others are correct: the ancient religious texts *are* crucial to a proper understanding of the Pyramid's ultimate purpose and function.

Sitchin's texts and his analysis are now reproduced here in the order of his presentation.

1. General Weapons Properties of the Great Pyramid

We learn more of the last phases of this Pyramid War from yet another text, first pieced together by George A. Barton (*Miscellaneous Babylonian Texts*) from fragments of an inscribed clay cylinder found in the ruins of Enlil's temple in Nippur.

As Nergal joined the defender of the Great Pyramid ("the Formidible House which is Raised Up Like a Heap"), he strengthened its defenses through various *ray emitting crystals* (mineral "stones") *Positioned within the Pyramid.*

> "TheWater-stone, the Apex-stone,
> The…-srtone, the…
> …the Lord Nergal
> increased its strength.
> The door for protection he… to Heaven its Eye he raised,
> Dug deep that which gives life…
> …in the house he fed them food."[8]

It should bc notcd that the interpretation of these magic "stones" as "ray-emitting crystals" is Sitchin's own. We shall encounter a different understanding of what these stones may have been with Christopher Dunn's version of the Machine Hypothesis (in a subsequent chapter).

2. Its Apparently Radioactive or Strong Electromagnetic Field Properties

Ninurta was at first astounded by (Ninhursag's) decision to "enter alone the enemyland"; but since her mind was made up, he provided her with "clothes which should make her unafraid" (*of the radiation left by the beams?*). As she neared the Pyramid, she addressed Enki: "She shouts to him… she beseeches him." The exchanges are lost by breaks in the tablct; but Enki agreed to surrender the Pyramid to her:

> "The House that is like a Heap,
> That which I have as a Pile Raised Up—
> Its mistress you may be."

---

[8] Sitchin, op. cit., pp. 163-164, emphasis added.

> There was, however, a condition: the surrender was subject to a final resolution of the conflict until "the destiny-determining time" shall come. Promising to relay Enki's conditions, Ninhursag went to address Enlil.[9]

Again, the conclusion that Ninhursag wore radioactive protective clothing is a conclusion that Sitchin himself makes, based upon his examination of this and numerous other texts. As will be discovered in examining Dunn's Machine Hypothesis, it is a conclusion *warranted* by some of the likely purposes of the structure itself.

3. The Motivation of the War

> Nowadays the visitor to the Great Pyramid finds its passages and chambers bare and empty, its complex inner construction apparently purposeless, its niches and nooks meaningless.
>
> It has been so even since the first men had entered the Pyramid. But it was not so when Ninurta entered it—circa 8670 B.C. according to our calculations. "Unto the radiant place," yielded by its defenders, Ninurta had entered, the Sumerian text relates. And what he had done after he entered changed not only the Great Pyramid from within and without but also the course of human affairs.
>
> When, for the first time ever, Ninurta went into "the House which is Like a Mountain," he must have wondered what he would find inside. Conceived by Enki/Ptah, planned by Ra/Marduk, built by Geb, equipped by Thoth, defended by Nergal, what mysteries of space guidance, what secrets of impregnable defense did it hold?[10]

---

[9] Sitchin, op. cit.., p. 165, emphases added.

[10] Sitchin, op. cit., p. 165. In the original footnote in *The Giza Death Star* I commented as follows: "It should be noted that one component of Sitchin's wider 'extraterrestrial Model' now intrudes. For Sitchin, the Pyramid also functioned as some sort of communications device guiding "ancient astronauts" to a spaceport in Sumeria. He wavers back and forth between these two functions," i.e., the communications

4. The Focus of Ninurta's Interest

A straight descending passage led to the lower service chambers where Ninurta could see a shaft dug by defenders in search for subterranean water.[11] But his nterest focused on the upper passages and chambers; there, *the magical "stones" were arrayed—minerals and crystals: some earthly, some heavenly, some the likes of which he had never seen. From them there were emitted the beamed pulsations for the guidance of the astronauts and the radiations for the defense of the structure.*

Escorted by the Chief mineralmaster, Ninurta inspected the array of "stones" and instruments. As he stopped by each of them, he determined its destiny—to be smashed up and destroyed, to be taken away for display, or to be installed as instruments elsewhere. We know of these "destinies", and of the order in which Ninurta stopped by the stones, from the text inscribed on tablets 10-13 of the epic poem *Lugal-e*. It is by following and correctly

---

and the weapons functions, "though his *texts* clearly indicate its primary function is as a weapon. Sitchin offers no explanation of how it might possibly have done both. Both are possible, particularly if it was a certain *type* of weapon embodying a certain *type* of physics." Suffice it to say here that this is a crucial point: that one and the same structure possessed both a communications (and therefore a broadcast and reception) function, and a weapons function.

[11] In the original annotated footnote in *The Giza Death Star* I commented as follows: "This requires some comment. Sitchin is doubtless referring to the so-called "Well Shaft" leading from the Grand Gallery. This notion strains credibility, because the inside chambers of the Pyramid, (even) during its operation in the relatively "benign" function Dunn ascribes to it, would have been uninhabitable, nor in its weapon mode would it have been necessary for its "defenders" to be located physically within it, any more than humans live inside nuclear reactors (or bombs). However, there is some merit to Sitchin's suggestion in that *if* the Weapon Hypothesis is true, the likelihood does exist of yet undiscovered (or undisclosed) chambers and passages far beneath the Giza complex. This idea is strengthened by the allegations of some, such as Herodotus, that such passages exist.

> interpreting this text that the mystery of the purpose and function of many features of the pyramid's inner structure can be finally understood.[12]][13]

At this juncture, Sitchin begins to elaborate more specifically what the text suggests the stones actually did.

5. The Textual Evidence Concerning the "Queen's Chamber"

> [Going up the ascending passage, Ninurta reached its junction with the imposing Grand Gallery and a horizontal passage. Ninurta followed the horizontal passage first, reaching a large chamber with a corbelled roof. Called the "Vulva" in the Ninhursag poem, this chamber's axis lay exactly on the east-west (axis) of the pyramid. Its *emissions ("an outpouring like a lion whom no one dares attack") Came from a stone fitted into a niche that was hollowed out in the east wall. It was the SHAM ("Destiny") stone. Emitting a red radiance which Ninurta "saw in the darkness," it was the pulsating heart of the pyramid. But it was anathema to Ninurta, for during the battle, when he was aloft, this stone's "strong power" was used "to grab to pull me, with a tracking which kills to seize me."* He ordered it "pulled out... to be taken apart... and to obliteration be destroyed."[14]][15]

a. Crystals as Histories of Local Space-Time Lattices

Here we must pause from our lengthy quotation of the original *Giza Death Star* text and its citation of Zechariah Sitchin's remarks, for the first of our extended revisions and extensions of remarks. Sitchin makes a number of quite important points in the previous paragraph, points which in turn form a rich matrix of

---

[12] Sitchin, op. cit., pp. 167-168, emphasis added.

[13] *The Giza Death Star: The Paleophysics of the Great Pyramid and the Military Complex at Giza* (Kempton, Illinois: Adventures Unlimited Press, 2001), pp. 44-49.

[14] Sitchin, op. cit., p. 168, emphasis added.

[15] *The Giza Death Star*, p. 50.

clues and concepts from which to extract the beginnings of basic physical principles possibly at work in the Pyramid. It is also important to note that at this juncture we are *not* concerned with the translational issues that many have raised with respect to Sitchin's translation and interpretation of the text; these issues will be dealt with subsequently. For the present, it suffices to observe that one way to distract people from the enormous implications of Sitchin's interpretation of the texts—that the Great Pyramid was a weapon—is to distract attention from that hypothesis by burying it in a blizzard of philological minutiae and details. With this in mind, we turn to that rich matrix of clues and concepts implied in this passage.

(1) Note first the reference to *the Sham or "Destiny" stone*. "Destiny" in the context of ancient texts, and particularly Mesopotamian texts as is the case here, implies an astrological or astronomical reference, i.e., *a particular configuration of celestial bodies, or, to put it in much more modern terms, a particular geometry or lattice structure of celestial bodies.* As every planetary or stellar body is of a particular mass, and in a particular location at a particular time, the overall *gravitational structure or lattice-work* of space-time is constantly changing. In other words, "destiny" should not be understood to mean "predetermination" but rather *structure* of that lattice work.

(2) Note secondly that this "sham" or "destiny" stone is a source of red radiation. We are not informed whether this is by reflection, or whether the stone *itself* is a source of this radiation, but the implication of the entire passage is that it is a radiation source, and not merely a reflection. Another implication is that it is a crystalline stone, and Sitchin himself takes the references to these "stones" as meaning "crystals."

(3) The third noteworthy thing, one confirming the view that the "sham" stone is a radiation source, is that it exerts a force in its vicinity, a "strong power" to "grab to pull me, with a tracking which kills to seize me." *These phrases strongly suggest that the force is not only strong, but either*

of electromagnetic, or gravitational nature, and possibly even both in some sort of electro-gravitic fashion.

(4) Finally, because of all of the previous points, Ninurta orders the "sham" or "destiny" stone to be destroyed, and here the language is again quite suggestive: it is to be "pulled out", i.e., presumably pulled out or removed from its special position in the Pyramid, and then it is to be "taken apart," *implying that the "stone" is composed of identifiable components or parts*, and finally it is to be destroyed "to obliteration.

Putting all this together, we get (1) a radiative crystal composed of (2) identifiable components exerting (3) a force of "destiny", i.e., an ability to manipulate the lattice structure of local space-time, and an electromagnetic, or gravitational, or electro-gravitic force *that (4) can target* a specific individual.

We may now extend the speculation.

Crystals are *grown*, and their lattice structure—including their lattice *defects—result from a response to the gravitational environment in which they are grown.* To take but a simple example, in very low gravity, crystals—depending on their chemical composition—may have more, or less, of certain types of lattice defects. Generally speaking, crystals grown in low or no gravity have relatively much fewer lattice defects than crystals grown in higher gravity. Thus, the types of lattice and lattice defects are dependent upon (1) the chemical and molecular structure of the atoms of the crystal itself, and (2) the local gravitational environment, and (3) the interaction between the two.

This implies that if one's science is sophisticated enough, one might be able to examine a crystalline lattice *and be able to deduce from that lattice the history of the local space-time conditions and its overall general trend of change, and from that in turn to make accurate predictions of the future of those conditions*. It may be possible, with such a science of "lattice-historiography" so to speak, to use such crystals to *manipulate* the wider environmental lattice of space-time, *for if crystal lattices are the result of the environmental conditions of local space-time over time, this means that those crystals also function as coupled harmonic oscillators of those conditions. The crystal, in effect, thus becomes an analogue observer of local space-time.*

It is now rather obvious why the general historical and predictive aspects of pyramid-as-prophecy hypotheses might have a general basis in physical mechanical truth.

With these "revisions and extensions" of remarks, we return to our quotation of Sitchin, and to our initial interpretation of them in *The Giza Death Star:*

> *[6. The "Grand Gallery" and Ninurta's Ascent*
>
> Returning to the junction of the passage, Ninurta looked around him in the Grand Gallery.... Compared to the low and narrow passages, it rose high (some twenty-eight feet).... Whereas in the narrow passages only "a dim green light glowed,"[16] the Gallery glittered *in multi-colored lights—"its vault is like a rainbow, the darkness ends there." The many-hued glows were emitted by twenty-seven pairs of diverse crystal stones that were evenly spaced along the whole length of each side of the Gallery. These glowing stones were placed in cavities that were precisely cut into the ramps that ran the length of both sides of the Gallery on both sides of its floor. Firmly held into place by an elaborate niche in the wall, each crystal stone emitted a different radiance, giving the place its rainbow effect.* For the moment Ninurta passed by them on his way up; his priority was the uppermost chamber and its pulsating stone.[17]

As will be seen, (Christopher) Dunn comes to a rather different conclusion as to what fit in the niches cut into the ramps (of the Grand Gallery), based solely on an engineer's examination of

[16] The original footnote here, note number 18, read "Ionized atmosphere gives off such a glow. And as we shall see, the Pyramid was filled with ionized hydrogen plasma." (*The Giza Death Star*, p. 51, n. 18)

[17] Zechariah Sitchin, *The Wars of Gods and Men*, Book III of *The Earth Chronicles* (New York, Avon Books: 1985), p. 168, emphasis added.

the pyramid. He believes that these slots or niches held banks of Helmholtz resonators.]

So I wrote as part of my original comment on this passage in the *Giza Death Star*. However, once again I feel it necessary to break my citation of the original book for a more extensive commentary than I originally made with respect to this passage from Sitchin.

It is clearly evident that Sitchin interpreted the Epic of Ninurta as an inventory of what was once the contents of the Great Pyramid. To put this point and its implications more clearly, Sitchin clearly understood the Pyramid to be a machine – a weapon – *but a machine whose most important components and parts had been removed from the present structure at some point in the remote past*, at some point *after* a war during which the Pyramid had been used as a weapon. The removal of the components was to prevent its use as such again.

Thus, what is present at Giza is not so much *a weapon*, but *the shell* of a weapon, like an old cannon in a public park whose barrel has been plugged and all the powder propellant and projectiles removed. Its most important components are missing, leaving us to speculate from the available clues what they and their underlying scientific principles might have been. By parity of reasoning, one cannot therefore understand the Pyramid nor its function, nor reconstruct any putative principles of its operation without a reference to relevant texts and a careful speculation on the kind of physics implied by those texts. They alone give hints of what the missing components might have been.

At this juncture, a question occurs which was never adequately answered in *The Giza Death Star*. The answer to this question was certainly implied in the original book, but never explicitly stated, because the question *itself* was left unstated. That question is: *Why* did Sitchin interpret the Epic of Ninurta and the *Lugal-e* in specific reference to the Great Pyramid in the first place? As will be seen, Sitchin himself answers this question, but only at the very end of his consideration of the Epic, and he answers it in such a way as to only *imply* the thought processes that led him to assume this approach to the text. Nonetheless, that answer, as will literally be seen, is clear and unequivocal.

We now return to the original *Giza Death Star's* examination of Sitchin:

> [As will be argued... in order to function effectively as a Weapon, the Grand Gallery would have had to incorporate some version both of Dunn's and Sitchin's apparatuses: artificial crystalline ... resonators that were both optically *and* acoustically resonant.

7. The "King's Chamber' and its "Coffer": The "Net" of Celestial Coupling

> (Ninurta) was now in the Pyramid's *most restricted ("sacred") chamber, from which the guiding "net" (radar?) was "spread out" to "survey heaven and earth" The delicate mechanism was housed in a hollowed out stone chest;*[18] *placed precisely on the north-south axis of the Pyramid, it responded to vibrations with bell-like resonance. The heart of the guidance unit was the GUG("direction determining") stone; its emissions, amplified by five hollow compartments*[19] *constructed above the chamber, were beamed out and up through two sloping channels leading to the north and south faces of the Pyramid. Ninurta ordered this stone destroyed: "Then by the fate-determining Ninurta, on that day was the GUG stone from its hollow taken out and smashed."*[20]

8. Commentary: The Parabolic Reflecting Faces of the Great Pyramid and a Common Mistake

At this juncture, it is necessary to interrupt the presentation of Sitchin's textual data once again in order to focus on what is a common misperception, and that is that the

---

[18] i.e., the "Coffer."

[19] Again, as will be shown, this is only partically correct, as the amplification effect was actually achieved by the huge granite stones that comprise the roofs of those **("relieving")** chambers, and not by the hollow chambers themselves.

[20] Zechariah Sitchin, *The Wars of Gods and Men* (Avon paperback version), p. 169, emphasis added.

"air shafts" leading from the "King's Chamber" (and by implication, the similar shafts leading from the "Queen's Chamber") served the function—for those who accept *any* version of the Machine Hypothesis (Weapon, Observatory, Communication or otherwise)—of emitting beams *from* the Pyramid outward. There is a property of the Great Pyramid, *unique to it alone of all the pyramidal monoliths on the Earth*, and that is that each of its four internal faces is indented slight at the center **(along the apothem, as was shown in chapter 1**); *beneath the flat casing stones, each face, in effect, constitutes a parabolic reflector much like a modern satellite dish.* As will be seen, only Christopher Dunn has correctly perceived that at least *one* of these shafts must be for the purpose of signal or energy *input*. This fact has a profound implication for the Weapon Hypothesis and to the interpretation of the rest of the Giza complex:

(1) parabolic reflectors are designed primarily to *receive, collect, and amplify signal input, not to send or transmit signals.* This implies that *one primary function of the Great Pyramid is as a collector and amplifier.* The question is, what kind of signals are being collected and amplified, and for what purpose?

(2) Satellite dishes and radio telescopes have an amplifier fixed at the focal center **(of the parabolic dish)** which collect and amplify the reflected signals. This has further implications:

  (a) part of the missing structural components of the pyramid either at one time resided *outside* the structure at some focal point in front of one, (or most likely, all) of its faces in a manner similar to a signal collector on a satellite dish. This would imply that some trace of such external structures as once may have existed should be evident, or may once have been evident, in the complex; ***or***,

  (b) those missing structural components still exist in other structures of the complex, but their function has not yet been adequately perceived; ***or,***

(c) the four faces of the Pyramid **(embody)** an advanced engineering principle *making the Pyramid itself that "collector and amplifier"—the view that will be defended subsequently—and further suggesting that none of the "air shafts" served as the primary output of whatever signal the Pyramid emitted, such output being the function, perhaps, of the missing apex "stone."*

A common question that is asked me whenever I have talked publicly about the Pyramid as a weapon is where its "signal" was emitted or "how was it aimed?" But this involves a massive misperception of the type of weaponry and physics that I am arguing, for it assumes that the Pyramid was some sort of "directed energy" weapon, like a particle beam or a laser, **(to be pointed and aimed at a target),** a view strengthened by Dunn's very cogent and persuasive argument that the "King's Chamber" employed a maser **(in its operation)**. This was *not,* however, the type of physics or weaponry **(that I am arguing was involved, for the weapon was not so much "pointed and aimed" at a target, but rather, *tuned to and brought into resonance with it, by a kind of non-local harmonic resonance).*** This means that its primary energy output came in the form of *non-linear* "tuned" energy. With this in mind, we may resume our survey of Sitchin

9. The "Grand Gallery" and Ninurta's Descent

> "Now came the turn of the minderal stones and crystals positioned around the ramps in the Grand Gallery. As he walked down Ninurta stopped by each one to declare its fate. Were it not for breaks in the clay tablets on which the text was written we would have had the names of all twenty-seven of them; as it is only twenty-two names are legible. *Several of them Ninurta ordered to be crushed or pulverized;* others, which could be used… were ordered given to Shamash; and the rest were carried off to Mesopotamia, to be displayed in Ninurta's temple in Nippur, and elsewhere, as constant evidence of the great victory….

All this, Ninurta announced, he was doing not only for his sake, but for future generations, too: *"Let the fear of thee"—the Great Pyramid—"be removed from my descendents; let their peace be ordained."*[21]

10. The Capstone

"Finally there was the Apex Stone of the Pyramid, the UL ("High as the Sky") stone: "Let the mother's offspring see it no more," he ordered. And, as the stone was sent crashing down, "Let everyone distance himself," he shouted. The "stones," which were anathema to Ninurta, were no more.[22]

This is an important piece of information, for it indicates that whatever else the missing Apex "stone" may have been, it was crucial to the function of the Pyramid *as a Weapon*....

11. The Victory Commemoration Seal

"The Second Pyramid war was over, but its ferocity and feats, and Ninurta's final victory at the Pyramids of Giza, were remembered long thereafter in epic and song—and in a remarkable drawing on a cylinder seal, showing Ninurta's Divine Bird within a victory wreath, soaring in triumph above the two Great Pyramids.[23]][24]

---

[21] Sicthin, *The Wars of Gods and Men* (Avon paperback edition), p. 171, emphasis added.

[22] Sitchin, *The Wars of Gods and Men* (Avon paperback edition), p. 171.

[23] Sitchin, *The Wars of Gods and Men* (Avon paperback edition), pp. 171-172.

[24] Joseph P. Farrell, *The Giza Death Star*, pp. 51-55.

Ninurta's Victory Seal

What is both extraordinary and necessary to note about this seal, which Sitchin has interpreted as a kind of commemorative "victory" seal, much like the painting of Bonaparte's victory over the Mamalukes at the Battle of the Pyramids, is that the pyramids depicted *are not the type of "stepped" or "terraced" pyramid or ziggurat that one would normally see in ancient Mesopotamia, where pyramids were "stepped" much like the Mexican pyramids, in discrete and easily seen "levels" or "stories", rather than the smooth-faced pyramids of Egypt.* This seal is in my opinion the key to Sitchin's process of reasoning, and the key that led him to interpret many of the texts which he cites as referring to the Great Pyramid, and its sister large pyramid, the Second Pyramid or Pyramid of Cephren. This is thus a significant clue as to when Sitchin thinks the "Second Pyramid War" was fought, namely, sometime *after* the Second Pyramid was constructed.

12. The Destruction of "the Great Weapon"

There was one final bit of text and quotations in the original *Giza Death Star:*

> ["After Ninhursag had finished her oracle of peace, Enlil was the first one to speak. *"Removed is the Affliction from the face of the earth,"* Enlil declared to Enki; *"the Great Weapon is lifted up."*[25]][26]

---

[25] Sitchin, *The Wars of Gods and Men* (Avon paperback edition), p. 174, emphasis added.

[26] Joseph P. Farrell, *The Giza Death Star*, p. 56.

With this last quotation, my original review of Sitchin's texts in *The Giza Death Star* was concluded, and notably, this last quotation, along with the references to *ekurs* or mountains/pyramids, to the "lower land" (Africa), and the "victory seal" clearly indicated that the structure was not only regarded as "the Great Weapon," but also regarded with some horror.

C. Sitchin and a Potential Rockefeller Connection

The high strangeness surrounding the Great Pyramid does not stop with Colonel Vyse's forgery and fraud, nor even with the fearsome allusions to weaponry in Sitchin's texts. Like Vyse, there is perhaps something peculiar about Sitchin himself that must be mentioned. While famous for being a scholar of ancient texts and for the bold hypothesis of his *Earth Chronicles* series based on his life-long hobby of examining those texts, one may have noted not only an uncanny familiarity with those texts, but also an ability to synthesize a vast quantity of them—and the scholarly literature *about* them—that one wonders if in fact his work and scholarship were being financially supported by hidden sponsors, and if in fact his books were not just for the general public, but more importantly, if they were really initially reports to those hidden masters, for howsoever one might disagree with this or that detail of his hypothesis of ancient high civilizations and wars, Sitchin's total *ouvre* remains an accomplishment and a synthesis scarcely since equaled in the field of alternative research.

By trade, Sitchin was an antique dealer in New York City, and given his eye for antiquities of a textual sort, one can only surmise that a similar talent to spot and trade valuable antiquities and antiques accompanied his plying of that trade.

It would have had to, for Sitchin's offices were in fact located in Rockefeller Center in New York City, and that, perhaps, is a clue to whom his possible hidden backers, if any, may have been.

D. The Tower of Babel Texts

Why associate the Tower of Babel with the Great Pyramid? The answer is rather simple, and in considering it, one is really confronting the idea that they are one and the same structure. In its most well-known and famous version, that presented in the eleventh chapter of Genesis in the Bible, the Tower of Babel is described as a "tower" that could "reach unto heaven" and that, moreover, allowed its builder, the first "world conqueror" Nimrod, and his fellow tower builders, to "do whatsoever they imagined to do." As we shall discover in a subsequent chapter, the Great Pyramid has many dimensional analogues to local celestial space, and as such, literally and metaphorically *is* a tower that "reaches unto the heavens. Indeed, until the construction of the Eiffel Tower in the nineteenth century, the Great Pyramid had remained for millennia the tallest structure on planet Earth.

1. The Biblical Version of the Story: The Tower's Cosmic Physics and Capability

In the version of the Tower of Babel story most readers of this book will be familiar with, the story occurs in the book of Genesis, chapter 11, verses 1-9:

> 1 And the whole earth was of one language, and of one speech.
> 2 And it came to pass, as they journeyed from the east, that they found a plain in the land of Shinar; and they dwelt there.
> 3 And they said one to another, Go to, let us make brick, and burn them thoroughly. And they had brick for stone, and slime they had for mortar.
> 4 And they said, Go to, let us build us a city, and a tower, whose top *may reach unto heaven;* and let us make a name, lest we be scattered abroad upon the face of the whole earth.
> 5 And the LORD came down to see the city and the tower, which the children of men builded.
> 6 And the LORD said, Behold, the people is one, and they have all one language; and this they begin to do: *and now nothing will be restrained from them, which they have imagined to do.*
> 7 Go to, let us go down, and there confound their language that they may not understand one another's speech.
> 8 So the LORD scattered them abroad from thence upon the face of all the earth and they left off to build the city.
> 9 Therefore the name of it is called Babel; because the LORD did there confound the language of all the earth: and from thence

> did the LORD scatter them abroad upon the face of all the earth.

In my book *The Tower of Babel Moment: Lore, Language, Leibniz and Lunacy* I commented as follows:

[As I have noted on a few occasions when talking about this story, there are a number of things that make this a very odd story for the Old Testament, for normally when "the LORD" intervenes in the affairs of men, or later, in the affairs of Israel elsewhere in the Old Testament, there is usually conveyed the idea that such intervention is required because of man's sinfulness. Indeed, the Flood itself in the biblical version was triggered for this reason, and one does not have to read very far into the Old Testament prophets to discover that divine interventions are usually responses to human iniquity of some sort.

That makes the *omission* of any such mention in the Tower of Babel story the more intriguing and significant. Of course, biblical commentators have in the main assumed that the pattern which is true *elsewhere* is true *here*, and have thus assumed that the divine intervention in this case was similarly motivated.

But read on its own and without reference to the rest of the Old Testament, the story suggests no such thing; considered in itself and *by* itself, the story does not explicitly *say* that. Rather, as I've pointed out in various talks on the story, the text alludes to something rather different: a linguistically-unified humanity is engaged in building some sort of structure—a "tower"—which can "reach unto heaven" which, in the words of "the LORD" himself, would allow men to do whatever "they have imagined to do" and to do it "without restraint." I have speculated that perhaps if one views the story as a metaphor, then the "structure" or "tower" which "reaches unto heaven" indicates perhaps a project designed to harness the *physics* of the creative heavens, and therewith great power, the power to do "whatever they imagine to do."

Interestingly, there *is* in chapter 10 of Genesis a brief mention that tends to confirm this hypothesis that the Tower represented a project to attain "heavenly" power, for according to the biblical text the founder of Babel (the city) was the first world conqueror, Nimrod. Citing chapter 10, verses 8-10:

8 And Cush begat Nimrod: he began to be a mighty one in the earth.
9 He was a mighty hunter before the LORD; wherefore it is said, even as Nimrod the mighty hunter before the LORD.
10 And the beginning of his kingdom was Babel, and Erech, and Accad, and Calneh, in the land of Shinar.

For many biblical commentators, the Tower of Babel Story is understood to be a gloss on, or continuation of, the story of Nimrod. In any case, from the biblical point of view, the founding of Babel by the "first conqueror" suggests the idea that the "Tower" was designed to continue those conquests to heaven itself, i.e., to the conquest and use of heavenly abilities.][27]

As has been observed previously, and as will be seen subsequently in this book, the Great Pyramid uniquely qualifies as a "tower reaching unto heaven," for in spite of not lying "in the plains of Shinar," it nonetheless incorporates many dimensional analogues of local celestial space. In this sense it "rcachcs" heaven, and can grasp the power of space, that is to say, the creative (and simultaneously, very destructive) power of the vacuum, of "the heavens", itself.[28] The third book of the Sibylline Oracles confirms this view of the Tower by stating that its builders "wanted to go up to starry heaven."[29]

2. The Slavonic and Greek Texts of Third Baruch: Boring into Heaven for Space-Warps

The Third Apocalypse of Baruch also contains glosses and elaborations of the Tower of Babel story. Before examining these, our assumption is that these stories, while occurring much later than the extant Old Testament or Mesopotamian texts and versions, nonetheless preserve and record genuine and authentic traditions. [The Slavonic version of chapters 2 and 3 of Third Baruch is as follows:

---

[27] Joseph P. Farrell, *The Tower of Babel Moment: Lore, Language, Leibniz, and Lunacy* (Lulu Books, 2020), pp. 15-16.

[28] Q.v. Joseph P. Farrell, *The Tower of Babel Moment*, p. 19.

[29] Joseph P. Farrell, *The Tower of Babel Moment*, p. 21, citing the Sibylline Oracles, 3:97-109.

2: 1 And the angel of hosts took me and carried me where the firmament of heaven is.
2. And it was the first heaven, and in that heaven he showed me very large doors. And the angel said to me, "Let us pass through these doors." And we passed through like the passing of 30 days. He showed me salvation.
3. And I saw a plain: *there were men living there whose faces were those of cattle, with the horns of deer, the feet of goats, and the loins of rams.*
4. And I Baruch asked the angel, *"Tell me what is the thickness of the heaven which we have crossed, and what is the plain, so that I can tell the sons of men."*
5. *Phanael said to me: "The gates which you saw are as large as (the distance) from east to west; the thickness of heaven is equal to the distance from earth to heaven, the plain where we are standing is equal to its width (i.e. heaven's)."*
6. He said to me, "Go and I will show you the mysteries."
7. I said to the angel, "Lord, who are these strangely shaped creatures?" And the angel said to me, "These are those who built *the tower of the war against God.* The Lord threw them out."

3:1. And the angel took me and led me to the second heaven and showed me large open doors, and the angel said to me, "Let us pass through them."
2. And we passed through, flying like the passing of 7 days.
3. *And he showed me a great prison, and there were strangely shaped creatures living in it, with the faces of dogs, the horns of deer, and the feet of goats.*
4. *And I asked the angel of the Lord, "Who are these?"*
5. *And he said to me, "These are the ones who had planned to build the tower,* for at that time they forced men and a multitude of women to make bricks. Among them was one woman who was near to giving birth, and they did not release her, but, working, she gave birth, and took her cloak and wrapped the infant, and left her infant, and made bricks again.
6. And the Lord God appeared to them and confused their languages. And they built their tower 80 thousand cubits in height, and in width 5 hundred and twenty.

7. *They took an auger so that they could proceed* ***to bore heaven so that they could see whether heaven is (made) of stone or of glass or of copper.***
8. And God saw them and did not heed them, but chastened them invisibly.[30]

The Greek text, while varying in minor points, is equally suggestive:

2:1. And taking me, he led me to where the heaven was set fast and where there was a river which no one is able to cross, not even one of the foreign winds which God created.

2. And taking me, he led me up to the first heaven, and showed me a very large door. And he said to me, "Let us enter through it." And we entered as on wings about the distance of 30 days' journey.
3. *And he showed me a plain within the heaven. And there were men dwelling on it with faces of cattle and horns of deer and feet of goats and loins of sheep.*
4. *And I Baruch asked the angel, "Tell me, I pray you, what is the thickness of this heaven in which we have journeyed, and what is its width, and what is this plain, that I may report these to the sons of men."*
5. And the angel, whose name was Phamael, said to me, *"This door which you see is (the door) of heaven, and (its thickness) is as great as the distance from earth to heaven, and the width of the plain which you saw is the same (distance) again."*
6. And again the angel of hosts said to me, "Come and I will show you greater mysteries."

---

[30] Citing H.E. Gaylord, Jr., trans., "Third Apocalypse of Baruch", in Hames H. Charlesworth, ed. *The Old Testament Pseudepugrapha: Volume 1: Apocalyptic Literature and Testaments* (New Haven: Yale University Press, 2009), p. 664, all emphases added. The scholarly debate on the dating of the text generally ascribes it to the first or second century A.D., with a possible reference to it in the early Christian writer Origen, giving a terminus post quem of ca. 230 A.D. (q.v. pp. 653-656).

7. *And I said, "I pray you, show me what those men are." And he said to me, "These are the ones who built the tower of the war against God,* and the Lord removed them."

3:1. And taking me, the angel of the Lord led me to a second heaven. And he showed me there a door similar to the first. And he said, "Let us enter through it."
2. And we entered, flying about the distance of 60 days' journey.
3. *And he showed me there also a plain, and it was full of men, and their appearance was like (that) of dogs, and their feet (like those) of deer.*
4. *And I asked the angel, "I Pray you, lord, tell me who these are."*
5. *And he said, "These are the ones who plotted to build the tower. These whom you see forced many men and women to make bricks.* Among them one woman was making bricks in the time of her delivery; they did not permit her to be released, but while making bricks she gave birth. And she carried her child in her cloak and continued making bricks.
6. And appearing to them, the Lord changed their languages; *by that time they had built the tower 463 cubits (high).*
7. *And taking an auger,* ***they attempted to pierce the heaven, saying, 'Let us see whether the heaven is (made) of clay or copper or iron.'***
8 Seeing these things, God did not permit them (to continue), but struck them with blindness and with confusion of tongues, and he made them be as you see.[31]

There are a number of very important points to note about these two versions of the Tower of Babel story.

---

[31] H.E. Gaylord, Jr., trans., "Third Apocalypse of Baruch", in James H. Charlesworth, ed., *The Old Testament Pseudepigrapha: Volume 1: Apocalyptic Literature and Testaments* (New Haven: Yale University Press, 2009), p. 665, all emphases added.

Firstly, it is crucial to observe that this is the first time that the story makes mention of *cruelty* with respect to the project, for both versions contain the poignant detail that a pregnant woman near delivery was forced to make bricks for the tower, during which she delivered her child, and then was forced to continue making bricks.

This is tied to the second detail, made clear in the Greek version of the text, that *two classes* of people were involved in the Tower project: (1) those who plotted and planned it, and (2) those who were forced to construct it, whereas previous versions of the story gave the impression that the Tower was freely planned and constructed by everyone.

The third intriguing detail common both to the Slavonic and (to the) Greek texts is the reference to "chimerical men" having the faces or feet of sheep, goats, gods, rams, and so on. There are three basic ways to interpret this: (1) the chimerical descriptions are meant as metaphors of the character of the people who plotted and planned the Tower project; or (2) the chimerical descriptions are intended to convey that those plotters and planners really *were* chimerical blends of different species, which implies the existence of a genetic technology able to accomplish that feat; or (3) both….

Fourthly, in both texts there are curious, attention-arresting mentions of the "thickness of heaven," and to a vast plain, and in both versions, this "thickness of heaven" is said to be as far as the earth is from heaven and the thickness of the *gates* to heaven is as "far as east to west." Again, one may read these expressions as simply metaphors for "infinity", or one may read them in a far more literal manner. I have thus far proposed the idea that the repeated references in various versions of the Tower of Babel story regarding the goal of "reaching the heavens" represented an intention to master the cosmic **(and creative)** power of the heavens, and if this be the case, then the references to the "thickness" of heaven, or to "plains", or to the distance from east to west, are far more literal *astronomical references, specifically, to the solar system* ***(and as will be seen here presently, to an even deeper physics possibility).*** On this view, then, the "plain" referred to is the plain of the ecliptic of the planets in orbit around the Sun, and "east" represents *the orientation toward the Sun*, and "west"

The orientation toward the outermost planet ***(or better, any orientation away from the Sun)****...*This in turn suggests that the Tower "reaching unto heaven" was designed to manipulate the local celestial mechanics of the Solar System, including the Sun.

Here we must pause to consider the implications of these references for our developing "high octane" speculative hypothesis, for if this reading be true, then the Tower, "reaching unto the heavens," if it were designed to tap into and manipulate (**such**) forces, falls into a "Class Two" civilization effort on the Kardashev scale, or at least into a Class Two on the "Farrell Corollaries" to that scale. Kardashev was the Russian astronomer who devised a system of classification of any extraterrestrial civilizations mankind might encounter as it ventured forth from Earth, or conversely, as those civilizations might come here. A Class One civilization required, in his definitions, the energy of an entire planet to sustain itself. A Class Two civilization required the energy of an entire star, or solar system, to sustain itself, and a Class Three required the energy of an entire galaxy to sustain itself.

However, in my previous talks about the possibility of a "secret space program" and "breakaway civilization", I have offered "Farrell corollaries" based on an analysis that I suspect may have been made after World War Two **(by the various "deep states" of the great powers, and particularly by those of the United States and the Soviet Union.)...** As UFOs became the focus of intense scrutiny both by amateurs and by various governments' militaries after World War Two, I reasoned that they were confronted by a strategic dilemma, a dilemma made more acute *if* they factored in such ancient stories and myths such as the Tower of Babel in their considerations, for that story demonstrated the vast implications of an external "intervention" in human affairs. In their turn, those considerations forced a question on those analysts: How would one go about "demonstrating" human capabilities to whomever may have been occupying or sending the UFOs to Earth? How would one, in other words, *prevent* a "Second Tower of Babel Moment" of history, particularly if one suspected that civilizations of a Class One or Two type were behind them (since a Class Three type would be so advanced that it would hardly be interested in us)? While humanity may be on the cusp of becoming a Type One civilization, we were and are not there yet. I reasoned **(therefore)** that one would have to demonstrate the ability to *engineer technologies* that could manipulate *systems of* a planetary or stellar scale, which implies the abilities to manipulate energies on those scales **(without requiring such enormous energies for the sustenance of the civilization engineering them).** This is a subtle difference from Kardashev's original classification scheme, but one which, given the situation at the end of World War Two, were just within the feasible technological possibilities of humanity.

In any case, back to third Baruch, for at the very end of both versions of the story in the Greek and Slavonic versions, there is an even more suggestive and arresting statement, for the Slavonic states that the project's planners "took an auger" in order to "*bore* heaven" in order to determine the composition of heaven, whether it be "of stone or of glass or of copper." In the Greek version, the auger is used to *pierce* the heavens, again with the purpose of determining the material composition of heaven. An auger, of course, is a large screw-like shovel, a manual *boring* machine designed to literally screw and twist its way through matter. The imagery, from a physics point of view, is thus highly suggestive, implying that the Tower was more than a simple tower, but that some component of it was deliberately designed to "probe the thickness of heaven" in a *very different way* than the astronomical, in a way designed to *bore*, via torsion, down to the fundamental components of matter and energy themselves, **(and if one recall the Sibylline Oracle, to *reach the "starry heavens"*, that is, to travel to them)....]**[32]

In other words, *the "thickness of heaven" might refer to* an even deeper physics than the astronomical, but to something much more fundamental, namely, *the metric of the lattice work of space-time, and the mass-energy conversion factor needed to warp* it.

Or to put the same thing "country simple," the "thickness" of heaven might refer quite literally to the amount of a warp in space time.

---

[32] Joseph P. Farrell, *The Tower of Babel Moment*, pp. 24-20, all emphases added by me in that book.

[3. Pseudo-Philo: The Tower as a Psychometric Object and the Semantic Confusion

In Pseudo-Philo the Tower of Babel Story is dispersed in passages occurring at the beginning of chapter 6, and again at the beginning and end of chapter 7:

> 6:1. The all those had been separated and were inhabiting the earth gathered and dwelt together. *And migrating from the east, they found a plain in the land of Babylon; and settling there, each one said to his neighbor,* "Behold it will happen that we will be scattered every man from his brother and in the last days we will be fighting one another. Now *come, let us build for ourselves a tower whose top will reach the heavens, and we will make a name for ourselves* and a glory *upon the earth."*
>
> 2. *And they said, each to his neighbor,* "Let us take bricks and let **each of us write our names on the bricks and** *burn them with fire;* and whatever will be burned through and through will be used for mortar and brick. [33]
>
> 7:2. And when they had begun to build, God *saw the city and the tower that the sons of men were building, and he said, "Behold they are one people and have one language for all; but what they have begun to make,* neither the earth will put up with it nor will the heavens bear to behold it. And if *they are not stopped now, they will be daring in all the things they propose to do.*
>
> 3. *And* behold now I will divide up *their languages* and scatter them in all regions so that a man will not understand his own brother and *no one will hear the language of his neighbor*.[34]

---

[33] D.J. Harrington, trans., "Pseduo-Philo", in James H. Charlesworth, ed. *The Old Testament Pseudepigrapha: Volume 2: Expansions of the "Old Testament" and Legends, wisdom and Philosophical Literature, Prayers, Psalms, and Odes, Fragments of Lost Judeo-Hellenistic Works* (New Haven: Yale University Press, 2009), p. 310, italicized emphasis in the original, boldface emphasis added by me. Pseudo-Philo is generally regarded to have been written sometime in the first century A.D.

[34] Ibid., p. 312, italicized emphasis in the original.

> 7:5. Now when the people inhabiting the land had begun to construct the tower, God divided up their languages and changed their appearances, and a man did not recognize his own brother *and no one heard the language of his neighbor.* **And so it happened that when the builders would order their assistants to bring bricks, those would bring water; and if they demanded water, those would bring straw. And so the plan was frustrated,** *and they stopped building the city. And the Lord scattered them from there over the face of all the earth. And therefore the name of* that place *was called "Confusion," bevause there* God *confused* their languages *and from there he scattered them over the face of all the earth,*[35]][36]

In this strange elaboration of the Tower of Babel story one encounters a new element of the deep physics being implied, namely, that "the builders of the Tower, by inscribing their names on the bricks or components of its construction" were assenting to the intention behind its construction, and thus viewing the Tower as "a kind of psychometric object." Pressing this observation further,

> this act signified their individual assent to whatever formally specified intention accompanied the Tower's construction. By thus creating a psychometric object (the Tower itself) comprised of components which themselves were psychometric objects (the "bricks") the attempt was being made to create both a manifold intention of *individuals in union with a unified group intention*.[37]

This implies a physics that views consciousness and intention can effect a physical system via the observer effect on a macro- as well as quantum-scale.

E. Other Ancient Texts Regarding Mounds, Mountains, Pyramids, and Cosmic Wars

---

[35] Ibid., p. 313, italicized emphasis in the original, boldface emphasis added.

[36] Joseph P. Farrell, *The Tower of Babel Moment*, pp. 31-32.

[37] Joseph P. Farrell, *The Tower of Babel Moment*, pp. 32-33.

The genius of Sitchin's analysis of the *Lugal-e* and other ancient texts in his book *The Wars of Gods and Men* was that he "cracked the code" of how to read or interpret ancient texts regarding the association of "wars of the Gods" with "pyramids." The language of mythology is, as De Santillana and Von Dechind pointed out in their ground-breaking study *Hamlet's Mill*, a language of science and technology, and once one knows its analogical and topological roots, one can decipher the code and the underlying principles of its magic.

However, the texts which Sitchin analyses in *The Wars of Gods and Men* are not the only ones pertinent to the weapon hypothesis of the Great Pyramid, nor to the wider issue of ancient cosmic wars involving weapons of mass destruction. One does not have to read very far in Vedic literature to comprehend that such wars are almost a driving theme of that literature. But as I pointed out in *The Giza Death Star Destroyed* and my book *The Cosmic War: Interplanetary Warfare, Modern Physics and Ancient Texts*, there are many other such texts. These texts do two things:

1. The texts persistently associate pyramids or mountains with wars, and various types of weapons; and,
2. The texts persistently associate *ekurs,* pyramids, mountains, or mounds with the very notion of creation and the power of creation, and thus imply a particular type of physics.

With these two points in mind, we now turn to re-examine the texts referenced in *The Giza Death Star Destroyed*, and *The Cosmic War,* for these texts constitute a necessary context both to the weapon hypothesis, to the ancient wars of mass destruction hypothesis, and to the hypothesis that there was a physics enabling both.[38]

1. Egypt: The Edfu Temple Texts, the Primeval Mound, and the Primeval War

[38] For the review of these texts in my book *The Cosmic War*, q.v. pp. 139-203.

While my book *The Cosmic War: Interplanetary Warfare, Modern Physics, and Ancient Texts* was primarily concerned with the war itself, it nonetheless investigated the descriptions of the weapons by which it was fought in several ancient texts, and some of those texts and sections are reproduced here because they bear directly upon the Weapon Hypothesis of the Great Pyramid, and afford significant and profound clues to the physics that made it work. Other portions of the argument that were presented in *The Cosmic War* in relation to certain texts are merely summarized. Once again, our convention of brackets—"[]"—is used when citing whole portions of previous books, which as before includes the original footnote in their original locations, but obviously not with the same numeration.

One of the most important of the ancient texts examined in the original *Giza Death Star* trilogy and *Cosmic War* books are the so-called "Edfu temple texts."

[So obscure are the Edfu texts that the best source for their study remains a secondary source which quotes extensively from them. This is E.A.E. Reymond's *The Mythical Origin of the Egyptian Temple,* published by the University of Manchester in England in 1969. Another important source is revisionist author Andrew Collin's *Gods of Eden: Egypt's Lost Legacy and the Genesis of Civilization.* The latter book is relied on here primarily due to the relative ease with which it is publicly available, although Reymond's work will also be extensively cited for certain details.

The Edfu texts are found inscribed on the walls of the temple of Edfu. Collins notes that what remains of this temple was "begun in 237 BC, and yet it was not completed until 57 BC."[39] Since each temple had its own "building text" that summarized "the name, nature, ritual significance and sometimes even the contents of decoration of the particular room,"[40] then it is "possible to draw up an outline picture of the nature and significance of temple as a whole" by conflating these texts.[41] Like many other

---

[39] Andrew Collins, *Gods of Eden: Egypt's Lost Legacy and the Genesis of Civilization* (Rochester, Vermont: Bear and Co., 2002, ISBN 187918176-2), p. 173.

[40] E.A.E. Reymond, *The Mythical Origins of the Egyptian Temple* (Manchester, England: The Manchester University Press, 1969), p. 4.

[41] Reymond, *The Mythical Origins*, p. 4.

Egyptian texts, however, the Edfu texts clearly stated that they were based on extremely ancient antecedents, for "according to legends carved" on the stone walls of the temple, the current structure was a replacement for a much older temple

> Designed in accordance with a divine plan that "dropped down from heaven to earth near the city of Memphis. Its grand architects were, significantly, Imhotep—a native of Memphis and, of course, a high priest of Heliopolis—and his father Kanefer.[42]

The Edfu texts, in other words, claimed a very ancient provenance.

But the "jewel in Edfu's crown" (is) its "so-called Building texts which adorn whole walls in various sections of the existing Ptolemaic temple."[43] It is here that E.A.E. Reymond's work enters the picture, for as Andrew Collins observes, "she was one of the few people who seemed to have grapsed the profound nature of the Edfu texts and realized that they contained accounts of a strange world that existed in Egypt *during what might be described as the primeval age."*[44] While "the texts at Edfu are many and varied", Collins observes that it is almost a certainty

> That much of their contents was derived from several now lost works, with titles such as the *Specification of the Mounds of the Early Primeval Age*, accredited to the god Thoth, the *Sacred Book of the Early Primeval Age of the Gods* and one called *Offering the Lotus*. All of these extremely ancient works begin with the gradual emergence out of the Nun, the primeval waters, of a sacred island, synonymous with the primeval mound of Heliopolitan tradition. This event is said to have occurred during a time-frame spoken of by Reymond as the "first occasion"—her interpretation of the Egyptian expression *sep tepi*, or the First Time.[45]][46]

---

[42] Collins, *Gods of Eden*, p. 173.
[43] Collins, *Gods of Eden*, p. 174.
[44] Collins, *Gods of Eden*, p. 174, emphasis added.
[45] Andrew Collins, *Gods of Eden,* p. 174.

It takes little imagination to square these doctrines with modern physics conceptions such as the zero point energy, or Diract Sea, or vacuum flux (or whatever one wishes to call it), the limitless sea of undifferentiated energy *from* which all matter, from sub-atomic particles to planets, stars, galaxies and galactic clusters, emerges, with the sea of Nun being the vacuum itself, the primeval waters the first differentiations of that sea into the emerging "mound" or "island" of matter, and so on.[47]

[Around this primeval mound, which was known as "the Island of the Egg" was a "channel of water," and on the edge ot it was a "field of reeds" that constituted a kind of sacred domain where columns "referred to as *djed*-pillars" were erected for the domain's "first divine inhabitants."[48] These were led by a group of "Sages" who were in turn led by "an enigmatic figure called... simply "This One."[49] These Sages of "faceless forms were said to have been the seed of their own creation at the time when the rest of the world had not yet come into being."[50] Indeed, these Sages

---

[46] Joseph P. Farrell, *The Cosmic War: Interplanetary Warfare, Modern Physics, and Ancient Texts* (Kempton, Illinois: Adventures Unlimited Press, 2007, ISBN 978-1-931882-75-90), pp. 166-168.

[47] Q.v. the original *Giza Death Star*, pp. 56-58. In that book I also observed that physicist and systems kinetics specialist Dr. Paul LaViolette had noted that when one graphs Plato's dimensions of Atlantis, one ends with a graph that is the two dimensional stationary wave profile of a proton. Q.v. Paul A. LaViolette, *Beyond the Big Bang: Ancient Myth and the Science of Creation* (Park Street Press, 1995), p. 223. It is to be noted that this stationary wave profile is a compressional or longitudinal wave, with the proton itself consisting both of a central mound, or nucleus, and "rings" or lesser concentrations of matter, which, when graphed, looks like a typical attenuating sin wave the further one progresses from the "nucleus" or central mound. This suggests, among many other factors examined in the texts here, that the physics being passed down in these texts is at least as sophisticated as contemporary particle physics and quantum mechanics.

[48] Andrew Collins, *Gods of Eden*, p. 174.

[49] Andrew Collins, *Gods of Eden*, p. 174.

[50] Andrew Collins, *Gods of Eden*, pp. 174-175. **(This doctrine is closely analogous to the ancient Stoic and Christian doctrines that every specific created being has its own "seminal logos" or creative**

are said by the Edfu texts to have preceded the appearance of the standard Egyptian gods.[51]

a. The "Sound Eye"

This tranquil state of affairs does not continue for long, however, for

> The Edfu account... *alludes to some kind of violent conflict which brought to a close the first period of creation. An enemy appears in the form of a serpent known as the Great Leaping One.* It opposes *the sacred domain's divine inhabitants,* who *fight back with a weapon known only as the Sound Eye, which emerges from the island and creates further destruction on behalf of its protectors.* No explanation of this curious symbol is given, although Reymond felt it to be the centre of the light that illumines the island. *As a consequence of this mass devastation, the first inhabitants all die....* Death and decay are everywhere—a fact recorded in the alternative names now given to the Island of the Egg, which include the Island of Combat, the Island of Trampling, and, finally, the Island of Peace.[52]

These statements positively compel and require commentary.

Firstly, it is to be noted that sometime after their "self-creation" from the "primeval waters"... a cosmic and cosmically violent conflict, a *war*, erupts. Moreover, the similarity here to Tiamat, whose name likewise symbolizes not only a planet, but primeval waters in the Babylonian *Enuma Elish*, is not to be overlooked. In other words, the Edfu text may be obliquely referring to the destruction of Tiamat-Krypton, Van Flandern's first exploded planetary event that took place ca. 65,000,000 years ago....][53]

The late Dr. Tom Van Flandern is of course the retired U.S. Naval Observatory astronomer who revived the 19th century idea

---

"seed principle/reason", and that creation as a whole is comprised of such internal rational principles, or *rationes seminales*, or λογοι σπερματικοι.)

[51] Andrew Collins, *Gods of Eden*, p. 175.

[52] Andrew Collins, *Gods of Eden*, p. 175, emphases added.

[53] Joseph P. Farrell, *The Cosmic War,* pp. 168-169.

that there was a missing planet in the solar system between Mars and Jupiter, a large solid planet that the late eighteenth and early nineteenth astronomers gave the suggestive name "Krypton." They hypothesized the existence of this planet on the basis of an astronomical law known as the Titius-Bode Law, or more commonly, simply as Bode's Law. Utilizing the principle of orbital harmonics, this law can be used to predict approximate orbital locations of planets if the mass of a system's central parent star is known. Bode's law predicted that the solar system should have another planet in orbit around the Sun at the approximate location of the asteroid belt. Astronomers of the day were mystified that they could not find the missing planet, until the first asteroid was discovered in the early nineteenth century, with others that quickly followed. Astronomers of the time quickly concluded that these little "planetoids" were the remnants of the missing planet that had exploded long ago, a planet that they named Krypton. Van Flandern revived this hypothesis, and among the hypothesized mechanisms that he proposed for why the planet exploded, he included "deliberate action," i.e., a technology gone terribly wrong, or a war.

While the war and its odd connection to Mars are endlessly fascinating,[54] [it is even more illuminating to note what the Edfu texts have to say about the weapons used in this ancient conflict. The inhabitants of the sacred "Island of the Egg" fight back with a weapon called the "Sound Eye," a weapon which the Edfu texts state emerges *from the island in the very same waters, i.e., from the very same transmutative medium. This strongly suggests that the "sound" involved was longitudinal pressure waves or stresses in the medium itself. In short, the "Sound Eye" may be a crude metaphor for these very types of "electro-acoustic" longitudinal waves and stresses in the medium*; the "Sound" part of the "Sound Eye" referring to the "acoustic" part of "electro-acoustic", and the "Eye" part referring to the "electromagnetic" part. "Sound Eye" is thus a crude, but synonymous way of saying "electro-acoustic."][55] As will be seen later in this chapter, such waves are another way of

---

[54] Q.v. Joseph P. Farrell, *The Cosmic War*, pp.170-171, for the connection of the war to Mars.

[55] Joseph P. Farrell, *The Cosmic War*, pp. 171-172.

expressing gravitational waves, and the whole process itself is that of creation from the vacuum, a process which in the case of the "Sound Eye" is weaponized. "Viewed in the paleophysical sense, then, the Edfu texts are describing *a process of creation"* by means of longitudinal waves in the physical medium, or if one prefer, the lattice work of space-time, a process "that possesses 'all the power of the universe.'"[56]

b. The "Rostau" and the Missing "Something"

[Note the sequence that we now have:

1. Sages "create themselves" from the "primeval waters," a metaphor, perhaps, for a paleophysics of a transmutative aether or medium, **(for a process of creation via longitudinal waves in the medium).** These sages then occupy an "island" in the midst of water.
2. A "Great Leaping One", a serpent, goes to war with the divine inhabitants of the Island, who defend themselves with the "sound Eye" and destroyed themselves in the process.
3. The Island recedes back into the primeval waters, and again reemerges with new occupants, who appear to be of a different, warlike and bloodthirsty character, than its first inhabitants.

These observations suggest that ownership or dominion of the island changed hands, a point whose full significance will be appreciated in a moment.

Then,

> After an undisclosed period, rising waters again threaten the Island of Trampling, causing the original temple... to be damaged or destroyed. Yet then something curious occurs... *Wa and Aa are instructed by the God-of-the-Temple to enter within the enigmatically named Place-in-which-the-things-of-the-earth-were-filled-with-power, another name for the water-encircled island, and here they conduct "magic spells" which make the*

---

[56] Joseph P. Farrell, *The Cosmic War,* p. 173, emphasis in the original.

> *waters recede. To this end they appear to have used mysterious power objects,* named ***iht,*** "relics', *which are stored* **within** *the island.*[57]

Once again, the text is capable of a deeper understanding.

To see how, one must unpack the italicized statements one by one.

Firstly, the gods Wa and Aa are instructed to enter "the enigmatically named Place-in-which-the-things-of-the-earth-were-filled-with-power," that is, they are required to enter some sort of structure, to be in or in proximity to it, in order to accomplish whatever it was they were doing. This important point raises yet another component of the... physics that I believe was a crucial element of the ancient technology, and that is *consciousness*.[58]

Secondly, note that the Place-in-which-the-things-of-the-earth-were-filled-with-power" is another name for the island itself, i.e., it is now a physical location that can be "entered."

Thirdly, they enter it in order to make use of "mysterious power objects" that are themselves stored inside the island.

This is beginning to sound an awfully lote like the mysterious "power stones" **(or Tablets of Destiny)** of Mesopotamian tradition... and an awfully lot like Ninurta's entry into the Great Pyramid to inventory its contents, an event described in detail by Zechariah Sitchin in his *Wars of Gods and Men* **(as we have already seen).**][59]

[It is Collins himself, however, who makes the identification of the "sacred island" of the Edfu texts with the Giza compound.

The key to this identification is to be found in the sacred islamd's title "Underworld of the Soul," a title given to it after the primeval conflict between the serpent, the "Great Leaping One," and its original inhabitants: "The term 'underworld' is the same here as that used in the Heliopolitan texts to describe the journey of

---

[57] Andrew Collins, *Gods of Eden*, p. 177, italicized emphasis added, bold and italicized emphasis in the original.

[58] Q.v. my *Giza Death Star Destroyed*, pp. 222-245.

[59] Joseph P. Farrell, *The Cosmic War*, pp. 174-175. Italicized emphases were added in *The Cosmic War*, simple boldface emphases are added here for explanatory clarity.

the sun-god through the *duat-underworld…*"[60] It is this journey that Osiris first undertook. And with Osiris comes the link to Giza, for in the famous Inventory Stela, discovered at Giza in thenineteenth century, reference is made to various structures on the plateau "including the Great Pyramid and the Great Sphinx." More importantly, it records "the visit to Giza of King Khufu." This is highly significant for it would appear that the ancient Egyptians were not in synch with modern standard Egyptology, since Khufu reigned before Khafre, the latter being the King standard Egyptology believes to have built the Great Pyramid!"[61]][62]

Actually, Collins has somewhat misrepresented the standard Egyptological construction of history, for it believes and attributes the Great Pyramid to Khufu *because he **initiated** its construction, while Khafre, or Cephren, completed it, and then went on to build the Second Pyramid of Giza.* But Collins' point is actually quite subtle, because he is actually maintaining that the "Inventory Stele" is exactly that, an *inventory of what was found and by whom, rather than of what was **constructed** and by whom.* [In other words, the Great Pyramid was **(actually being said)** by the ancient Egyptians themselves to be of greater antiquity than current standard theory allows.

But this is not all.

The Inventory Stela "also makes reference to the "House of Osiris, the Lord of Rostau." Since the Sphinx is cited as being 'northwest' of this 'House of Osiris,' it is thought by some writers to be a reference to the Valley Temple, which lies roughly east-south-east of the Sphinx monument."[63] Collins himself draws the inescapable conclusion: "It therefore seems certain" that the "sacred island" and home of the "Sound Eye" and all its environs "—with its field of reeds, water-encircled island and temple complex—*was almost certainly located at Giza."*[64]

---

[60] Andrew Collins, *Gods of Eden*, p. 179.
[61] Andrew Collins, *Gods of Eden*, p. 180.
[62] Joseph P. Farrell, *The Cosmic War*, p. 177.
[63] Andrew Collins, *Gods of Eden*, p, 180.
[64] Andrew Collins, *Gods of Eden*, p. 180, emphasis by Collins.

Giza, in other words, was directly associated with the "true Point of First Creation,"[65] referred to in the Edfu texts, as Collins puts it, it is "an attractive supposition." But if this is so, then it is also the true point of the "Sound Eye" and the focus and center of the terrible conflict with the "Great Leaping One" that brought an end to that first creation, and now, more bloodthirsty occupants of the "sacred island." It was, in short, the location of the weapon that brought a cataclysmic end to a conflict in the pantheon and ruin to the people that wielded it.

But Collins *does* mention another difficulty:

> There is, however, one major problem—how might we reconcile the appearance of the plateau today with the description of the sacred island) as a temple located beside a field of reeds, situated on the edge of a primeval lake containing a small mound-like island?[66]

Collins' answer is "that in the *eleventh millennium BC*, low-lying areas east of the Giza plateau were regularly, if not permanently, flooded to create a shallow lake that may have encircled a small rocky island."[67]][68]

In other words, the Inventory Stela is a case of "cultural appropriation" by Khufu(Cheops) and Khafre(Cephren), and the two pyramids, and the Sphinx, are millennia older than dynastic Egypt itself, and belong to an age or layer of construction that is congruent with the Sahara sub-pluvial period.

2. Sumer and Mesopotamia:
a. The Legend of Erra/Nergal and Ishum

Turning to Mesopotamia, there is another short but significant text that gives insight not only into the nature of the war between the gods, but also into the nature of the physics behind the weaponry that was deployed. As I noted in *The Cosmic War*, this

---

[65] Andrew Collins, *Gods of Eden*, p. 180.
[66] Andrew Collins, *Gods of Eden*, p. 180.
[67] Andrew Collins, *Gods of Eden*, p. 181, emphasis added by me.
[68] Joseph P. Farrell, *The Cosmic War*, pp. 177-178.

text should be read in conjunction with the *Enuma Elish*.[69] The text is *The Legend of Erra and Ishum*, and concerns a council of the gods, including the supreme god Marduk, and their efforts to persuade the god Erra to go into open revolt. Erra is yet another name for Nergal, god of wars, and particularly wars of rebellion. Nergal, as the god of war, is the name of the planet Mars in Mesopotamian mythology, and of course is synonymous with another of his Mesopotamian names, Errakkal or Herakkal, who becomes the Greek Hercules and the Roman Aires.

In the council of the gods, Nergal is counselled to go into total and open revolt and to seek absolute power:

> Go out to the battlefield, warrior Erra, make your weapons resound!
> Make your noise so loud that those above and below quake,
> So that the Igigi[70] hear and glorify your name,
> So that the Annunaki hear and fear your word,
> So that the gods hear and submit to your yoke,
> So that kings hear and kneel beneath you,
> So that countries hear and bring you their tribute,
> So that demons hear and avoid(?) you,
> So that the powerful hear and bite their lips.
> ....
> Warrior Erra listened to them.
> The speech which the Sebitti made was as pleasing to him as the best oil.[71]

[One god, Ishum, steps forward to council Nergal against open revolt and war:

> When Ishum heard this,
> He made his voice heard and spoke to the warrior Erra,
> "Lord, Erra, why have you planned evil for the gods?
> You have plotted to overthrow countries and to destroy their people, but will you not turn back?"[72]

---

[69] Ibid., pp. 149-150.

[70] "Igigi" s simply the term for men, or human beings.

[71] Stephanie Dalley, trans. and ed., *Myths from Mesopotamia: Creation, the Flood, Gilgamesh, and Others* (Oxford: Oxford University Press, 2000, ISBN 0-19-283589-0), pp. 287-288.

Undeterred, Nergal resolves to assault the king of the gods, Marduk himself, and actually confronts Marduk personally.

Marduk's response is to remind Nergal of his power, by reminding him how he destroyed Tiamat and *rearranged the very mechanics of heaven itself:*

> The king of the gods made his voice heard and spoke,
> Addressed his words to Erra, warrior of gods,
> 'Warrior Erra, concerning that deed which you have said you will do:
> *A long time ago, when I was angry and rose up from my dwelling and arranged for the flood,*
> *I rose up from my dwelling, and the control of heaven and earth was undone.*
> *The very heavens I made to tremble, the positions of the stars of heaven changed, and I did not return them to their places.*
> Even Erkalla quaked....
> *Even the control of heaven and earth was undone...*"[73]][74]

Marduk is referring to the events of a prior "cosmic war of the gods", his titanic struggle with the goddess Tiamat as it was recorded in the *Enuma Elish*. Here I will cite at length what I wrote about the *Enuma Elish* in *The Giza Death Star Destroyed*, and which I repeated in *The Cosmic War*.

[b. The Enuma Elish and the Broken Cosmic Order

...The Babylonian "creation epic", the *Enuma Elish*, presents a concise though quite suggestive account of this **(earlier)** interplanetary rebellion and war. The principal characters... are the "gods" Tiamat and Marduk.

The account begins as a creation account in rather typical fashion for an ancient text, recounting a state of initial chaos from which, through conflict of opposites, the order of creation gradually emerges:

---

[72] Ibid. p. 289.
[73] Dalley, *Myths from Mesopotamia,* p, 290, emphasis added.
[74] Joseph P. Farrell, *The Cosmic War*, pp. 149-150.

1. When in the height heaven was not named,
2. and the earth beneath did not yet bear a name,
3. and the primeval Apsu, who begat them,
4. *and chaos, Tiamat, the mother of them both—*
5. *their waters were mingled together,*
6. and no field was formed, no marsh was to be seen;
7. when of the gods none had been called into being;
8. *and none bore a name, and no destinies {were ordained}...*[75]

Note that the first state or condition of creation is an undifferentiated state, or "chaos", a condition recalling **(my "topological metaphor of the medium which I have written about in other books).**[76] This undifferentiated state would best be described by our modern physics terms of "vacuum", "zero point energy", "quantum flux" or even "medium" or "aether." The occurrence of the concept here, prior to the appearance of any distinctive objects of creation, is a strong indicator that the document preserves a residue of an earlier more sophisticated "paleophysical" cosmology. This is corroborated by the absence of names—"and none bore a name, and no destinies {were ordained}"—indicating in another fashion the absence of physically distinct and observable characteristics, corroborating the idea that we are dealing with a document of cosmological physics guised in a religious text.

This might suggest that the proper names for the "gods" would argue against the "titular" *pars pro toto rhetorical usage* advocated by Laurence Gardner **(as a way of understanding and interpreting the proper names of gods in ancient texts).** As will be seen, however, the association of "Tiamat" with chaos and destruction subsequently in the *Enuma Elish* may be an artifact of the role of actual persons in the war described subsequently in the epic. In fact, the indicator of this war occurs almost immediately after the opening verses cited above, strongly suggesting that the conflict was a very ancient one, **(occurring almost as soon as the**

---

[75] *Enuma Elish* ed. L.W. King, M.A., F.S.A., Vol. I I(London: luzac and Co., 1902), p. 3, Tablet 1 (the number of verses are from the edition cited).

[76] Q.v. Joseph P. Farrell, *The Philosophers' Stone* (Port Townsend, Washington: Feral House).

creation was completed, in a manner quite similar to the Edfu temple texts):

> 22. But T{iamat and Apsu} were (still) in confusion {…},
> 23. They were troubled and {……}
> 24. In disorder(?)..{……}
> 25. Apsu was not diminished in might {….}
> 26. and Tiamat roared {……..}
> 27. She smote, and their deads {……}
> 28. Their war was evil …{.}….

Notwithstanding the deteriorated condition of the tablets from which the text is translated, there are clear indications that Tiamat and Apsu were real persons, since they are engaged in activities perceived as *evil.* Soon after this, the epic gives the reason for this moral assessment:

> 49. Come, their way is strong, but thou shalt destroy {it};……
> 51. Apsu {hearkened unto} him and his countenance grew bright,
> 52. {Since} he (i.e. Mummu) planned evil against the gods his sons.[77]

Note again the personalism of the document, as well as the fact that the war appears to be a "family quarrel" that has erupted into a civil war, a reading well in line with Zehariah Sitchin's reconstructions **(and in line with the Judeo-Christian version of a revolt of Satan and a "war in heaven").** Many of the "gods" quickly flock to Tiamat's side.

At this juncture, the epic becomes very specific—unusually specific in fact, for a mere "creation epic"—in cataloguing the weapons used by the "Tiamat alliance"; **(in fact, it is at this juncture that the "creation epic" reveals itself to be in actuality a "cosmic war epic"):**

> 109. {They banded themselves together and} at the side of Tiamat {they} advanced;

[77] *Enuma Elish*, Tablet 1, p. 7.

110. {They were furious, they devised mischief without resting} night and {day}.
111. {They prepared for battle}, fuming and raging;
112. {They joined their forces} and made war,
113. {Ummu-Hubu}, who formed all things,
114. {made in addition} *weapons invincible, she spawned monster serpents,*
115. *{sharp of} tooth, and merciless of fang;*
116. *{with poison instead of} blood she filled {their} bodies.*
117. Fierce {monster-vipers } she clothed with terror….
…
120. Their bodies reared up and *none could withstand {their} attack.*[78]
121. {She set} up vipers, and dragons, and the (monster) {Lamamu},
122. {and hurricanes}, *and raging bounds, and scorpion-men,*
123. *and mighty {tempests}, and fish-men, and {rams};*
124. {They bore} cruel weapons, without fear of {the fight}.[79]][80]

Now let us pause, and take note of the science implied by wielding the weapons described in this passage.

There are essentially two types of weapons that the "Tiamat Alliance" deploys and uses in its war: (1) biologically or genetically engineered life forms, and (2) weather weapons. With respect to the biological weapons, these are of two further sub-classes: (a) some are larger than normal size and are reptilian, strongly suggesting dinosaurs or similar creatures, a fact that in turn suggests the great antiquity not only of the wars, but that such creatures were engineered as weapons of war; and (b) some of these—e.g., "fish-men" and "scorpion-men"—are chimeras or hybrids between species, suggesting not only an advanced technology, but the possibility that the artistic and hieroglyphic

---

[78] *Enuma Elish*, Tablet 1, p. 17, emphasis added.
[79] *Enuma Elish*, Tablet 1, p. 19, emphasis added.
[80] Joseph P. Farrell, *The Cosmic War*, pp. 150-153.

depiction of such creatures in Egyptian and Mesopotamian art and writing might rest on some very real but temporally remote antecedent. With respect to the second category of weapon, weather weapons, the engineering of hurricanes and other immense storms suggests minimally the ability to engineer systems on a planetary scale. But "hurricanes" might be understood to be a metaphor for much larger vorticular "storms" such as singularities like black holes or "mini" black holes.

[But what were the motivations for this titanic struggle and the development and actual *use* of such horrendous weapons? These are alluded to at the very end of the first tablet and again in the second tablet of the *Enuma Elish*:

> 137. *She gave to (Kingu) the Tablets of Destiny, on {his} breast she laid them, (saying):*
> 138. *"Thy command shalt not be without avail, and {the word of thy mouth shall be established}."*
> 139. *Now Kingu, (thus) exalted, having received {the power of Anu}...*[81]

The reference to "the power of Anu" is significant, since "Anu" in the Babylonian theogony is the name of the supreme God. Thus, whatever the "tablets of Destiny" were, they conveyed such tremendous power to their possessor that the power was regarded as being divine, **(indeed, perhaps the power of creation and destruction, and the power to be "present everywhere").**

....

But what was the reason for this war? According to the epic, it was Tiamat's possession of these mysterious "Tablets of Destiny" and the extreme power they conveyed. This constituted the primary reason for the war against her and for her utter destruction at the hands of her opponents. And this brings us back to Marduk, who was the chief of her opponents in the epic, and to *his* arsenal and to "The Sequence of Tiamat's Destruction."

(1) The Sequence of Tiamat's Destruction, Marduk's Weaponry, and the Science Implied by Them

---

[81] *Enuma Elish* Tablet 1, 0. 21, emphasis added, q.v. also Tablet 2: vv. 43-45, p. 29.

The epic quickly moves to the topic of the appointment of Marduk as the leader of the coalition to defeat Tiamat:

> 13. O Marduk, thou art our avenger!
> 14. We give thee sovereignty over the whole world.
> 15. Sit thou down in might, be exalted in thy command.[82]

Marduk a little further on is then given a very interesting mission:

> 31. Go, and cut off the life of Tiamat,
> 32. and let the wind carry her blood into secret places.[83]

The significance of this mission will be lost unless one bears in mind the titular *pars pro toto* paradigm **(where the proper names of gods not only designate real persons or sovereigns but also the celestial—the planet—they govern).** On that view Marduk is charged to destroy the entire planet represented by the titular term "Tiamat." The horrendous biological and weather weapons Tiamat has unleashed on her opponents has called forth an escalation of the war as her opponents now call for her complete destruction….

In this same context, Marduk is then given two rather interesting weapons, from this "paleophysical" point of view, one of which, perhaps, represents some form of stealth technology:

> 23. "Command now and let the garment vanish,
> 24. and speak the word again and let the garment reappear!"
> 25. Then he spake with his mouth, and the garment vanished,
> 26. Again he commanded it, and the garment reappeared.[84]

In addition to this "Stealth suit," Marduk is given an "invincible weapon," a weapon far exceeing Tiamat's biological and weather arsenal:

---

[82] *Enuma Elish*, Tablet 4, p. 59.
[83] *Eunma Elish*, Tablet 4, p. 61.
[84] *Enuma Elish*, Tablet 4, p. 61.

27. When the gods, his fathers, beheld (the fulfillment of) his word,
28. they rejoiced, and they did homage (unto him, saying) "Marduk is King!"
29. They bestowed upon him the scepter, and the throne, and the ring,
30. *They have him an invincible weapon, which overwhelmeth the foe.*[85]

What this invincible weapon may be is not described nor named, but its *effects* **(are catalogued)** in the following passage:

39. *He set the lightning in front of him,*
40. *with burning flame he filled his body,*
41. *he made a net to enclose the inward parts of Tiamat,*
42. *the four winds he stationed so that nothing of her might escape;...*
43. The South wind and the North wind and the East wind and the West wind
44. *He brought near to the net, the gift of his father Anu.*
45. *he created the evil wind, and the sevenfold wind, and the whirlwind, and the wind which hath no equal.*[86]

This passage implies a rather remarkable set of characteristics of the "invincible weapon":

(a) Its use apparently involved lightning, i.e., extremes of electrostatic energy;
(b) The "net" used to "enclose the *inward* parts of Tiamat" recalls the language of another Babylonian epic, the *Lugal-e*, which forms so much of the material Zechariah Sitchin used to reconstruct his "Second Pyramid War" **(as was covered earlier in this chapter. This terminology suggests)** a weapon employing *gravity and acoustics, i.e., longitudinal waves,* as its primary component.... Thus, Marduk's "invincible weapon" appears to be able to tap into the field of

[85] *Enuma Elish,* Tablet 4, p. 61, emphasis added.
[86] *Enuma Elish*, Tablet 4, p. 63, emphasis added.

space-time, i.e., the medium itself, and thereby the planetary *center* of Tiamat….

(c) If one understands the "net: in this fashion, as the gridwork or lattice of cellular-like structure of the medium itself, then further corroboration of this **(interpretive heuristic and hermeneutic)** would appear to be provided by the reference to the "four winds," which might be taken to mean the compass points, or even more abstractly, *coordinate references*. **(As will be seen subsequently in this chapter, a similar reference to a weapon tied to the four points of the compass, or the four "corners" of the Earth, occurs in ancient Hindu texts concerning the "war of the gods"]**[87]

There is, however, *another* interpretive possibility for understanding references such as "the four winds", or in some cases, "the four corners" of the Earth. It is well-known that the simplest solid that can be circumscribed in a sphere is a tetrahedron, a four-sided pyramid:

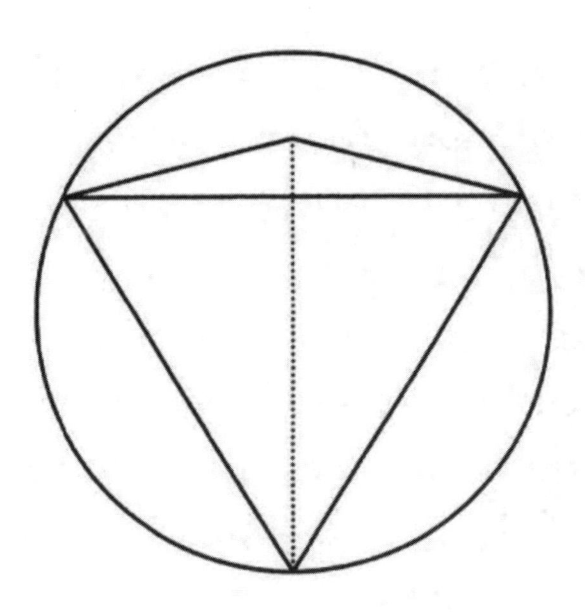

[87] Joseph P. Farrell, *The Cosmic War*, pp. 154-157.

Spherically Circumscribed Tetrahedraon

As will be seen in a later chapter, the Great Pyramid employs throughout its construction the principle of a "squared circle" and a "cubed sphere." As such, *ancient references to the "four winds" or to the "four corners of the earth" in connection with such weaponry* ***could be referring to actual pyramidal structures,*** *and as we shall discover, in particular to the Great Pyramid as an analogue and oscillator of the Earth itself.*

[With these thoughts in mind, we turn to the Sequence of Tiamat's Destruction itself. This "Sequence" comprises the *main theme* of the Fourth Tablet of the *Enuma Elish*. In it, as we shall see, the signature of Marduk's "invincible weapon" points very strongly to it being the type of "scalar" **(or "electro-acoustic" or "electro-gravitic")** weapon employing such distortions or **(longitudinal)** pressure waves in the medium as its primary component...:

> 47. *He sent fourth the winds which he had created, the seven of them;*
> 48. *to disturb the inward parts of Tiamat, they followed after him.*
> 49 Then the Lord raised the thunderbolt, his mighty weapon,
> 50. *He mounted the chariot, the storm unequalled for terror,*
> 51. *He harnessed and yoked it unto four horses,*
> 52. *Destructive, ferocious, overwhelming, and swift of pace...*[88]
> 58. With overwhelming brightness his head was crowned...
> ...
> 65. And the Lord drew nigh, he gazed upon the inward parts of Tiamat...[89]
> 75. *Then the Lord {raised} the thunderbolt, his mighty weapon...*
> 76. {and against} Tiamat, who was raging, thus he sent (the word):
> 77. "Thou art become great, thou hast exalted thyself on high,
> 78. and thy {heart hath prompted} thee to call to battle...."[90]
> ...
> 87. *When Tiamat heard these words,*

---

[88] *Enuma Elish,* Tablet 4, p. 65, emphasis added.
[89] *Enuma Elish*, Tablet 4, p. 67.
[90] *Enuma Elish,* Tablet 4, p. 69.

88. *She was like one possessed, she lost her reason.*
89. *Tiamat uttered wild piercing cries,*
90. *she trembled and shook to her very foundations....*

...

95. *The Lord spread out his net and caught her,*
96. *and the evil wind that was behind him he let loose in her face.*
97. As Tiamat opened her mouth to its full extent,
98. *He drove in the evil wind, while as yet she had not shut her lips.*
99. *The terrible winds filled her body,...*

...

101. *He seized the spear and burst her body,*
102. *he severed her inward parts, he pierced (her) heart.*[91]

Afterwards, Marduk captures Tiamat's allies,[92] recapturing the Tablets of Destiny from Kingu, and then returns to Tiamat to complete her destruction:

129. And the Lord stood upon Tiamat's hinder parts,
130. and with his merciless club he smashed her skull.[93]
131. *He split her up like a fish into two halves;...*
143. And the Lord *measured the structure of the Deep.*[94]

I believe these passages reveal a remarkably accurate sequence of what the destruction of a planet by a "scalar" weapon employing a longitudinal pulse or "acoustic"" or "gravitational" wave in the medium would entail, right down to the acoustic cavitation of a planet **(resulting in extreme earthquakes and noise to)** large electrostatic displays **(or lightnings)**, signatures of such a weapon at extreme power. Let us note the sequence:

(a) The "winds" **(i.e., the longitudinal pulses in the medium itself)** are sent to "disturb" or destabilize the "inward parts" of Tiamat, the planetary core (vv. 47-48);

---

[91] *Enuma Elish*, Tablet 4, p. 71, emphasis added.
[92] *Enuma Elish*, Tablet 4, vv. 105-112, p. 73.
[93] *Enuma Elish*, Tablet 4, p. 75.
[94] *Enuma Elish*, YTablet 4, P. 77, emphasis added.

(b) "Lightning" is then unleased on the already unstabilized planet from the "four winds", i.e., **(from an oscillator of an entire planet);**[95]

(c) These "thunderbolts" are then apparently directed toward that destabilized core, suggesting that a sudden and extreme *pulse* is administered (vv. 58, 65, 75-78);

(d) Tiamat respons with cries and trembles and shakes to "her very foundations", i.e., experiences very severe earthquakes or acoustic cavitations **(with accompany-ing noise)** throughout the planet, to its very core (vv. 95, 97f);

(e) Tiamat appears unable to break resonance with the weapon (vv. 97-98) as Marduk spreads the net and drives in the final "wind" or pulse (v. 98);

(f) Tiamat reaches maximum instability in her planetary core and mantle (vv. 98-99);

(g) Marduk pierces the crust, and releases enormous energies that have built up in the planet through the acoustic cavitations, resulting in a colossal explosion with the entire planet as its fuel, rather like bursting a balloon filled to extreme pressure. (vv. 101-102, 137).][96]

The *Enuma Elish* thus implies, at almost every step in its description of the war and of Marduk's "invincible weapon" a physics not only sufficient to measure "the structure of the deep" but to weaponize it via compression, or longitudinal waves in its lattice work. As one is compressing , that is to say, *warping* space time, anything caught in the region of that warp could literally be crushed, or so loaded with energy, that it explodes.

[It is to be noted that Marduk "measures the structure of the Deep" *after* Tiamat's destruction. This would have been necessary in terms of the type of physics being suggested, since the destruction of a planetary-sized body in the approximate orbit of the asteroid belt would have required an adjustment to astronomical measurements of the solar system, since its previously existing celestial mechanics and geometry have been

---

[95] This is a crucial clarification of the concept made here in this book; it does not appear in the original *Cosmic War*.

[96] Joseph P. Farrell, *The Cosmic War*, pp. 158-160.

shattered.][97] In other words, *the "wars of the gods" are not metaphors of natural recurring catastrophes such as comets or asteroids crashing into planets and causing devastation, as many have argued. Rather the reverse: a real war or wars are the origins of the cycle of catastrophe.*

(2) A Tangent into Ufology: Morris Jessup, Carlos Allende, The Varo Edition, "The Great Bombardment" and Kinetic Weapons: Hurling Asteroids and Thunderbolts

While it is not strictly necessary, it may nonetheless be worthwhile at this juncture to take a small detour into the field of Ufology, and to curious references to the ancient cosmic war that occurred in relationship to a book on UFOs and their implied science, written by the 1950s Ufologist and professional academic astronomer and amateur archaeologist, Dr. Morris K. Jessup. The book, *The Case foe the UFO: Unidentified Flying Objects* was published in by Bantam paperback books in 1955.

Jessup appeared on a few radio shows at the time that were willing to discuss UFOs, and somehow a man calling himself "Carlos Allende" either heard one of these shows, or had purchased a copy of Jessup's book. In any case, the U.S. Navy soon started receiving a series of letters from Allende—three in all—detailing allegations of a top secret wartime experiment conducted by the U.S. Navy trying to make a warship invisible to radar. The result, according to Allende's letters, was something completely unanticipated. The ship was not only rendered completely invisible, but was instantly teleported hundreds of miles. The crew, Allende alleged, suffered equally strange side effects, from being partially merged with the ship in its bulkheads, or later suffering spontaneous combustion or simply completely disappearing from sight, never to be seen again.

Allende was, of course, talking about the Philadelphia Experiment and the story he related to Jessup in his three letters has since become part of the core narrative of that story.

---

[97] Ibid., p. 161.

For our purposes, what interests us is what happened after this initial contact, for Allende subsequently sent a copy of Jessup's UFO book to the Navy, which copy contained heavy marginal notations in different colors of ink, with some of the notes supposedly by Allende and some by two other people. The Navy turned the annotated book over to the Navy's intelligence division, the Office of Naval Intelligence, which found the annotated version of the book so intriguing that it contacted the Varo Publishing Company of Garland, Texas to reproduce the entire book, complete with the annotations in colored ink. Approximately 200 copies were made and distributed to various people, among whom, as I argued in my book *Secrets of the Unified Field*, was the German rocket scientist Wernher Von Braun.[98] Another copy was given to Dr. Jessup himself, who consulted with the US Navy for a time on the topics of the notations, and who subsequently committed "suicide" under clearly mysterious and questionable circumstances…

…but not before giving copy of the Varo edition to a friend of his for safekeeping. It is this copy, I believe, that is the ultimate and original source for the Varo edition that has circulated in the alternative research "underground" ever since.

In the Varo edition of Jessup's book, the annotations refer repeatedly to an ancient and clearly interplanetary war that goes by various names, "the great battle,"[99] and the "great bombardment," which brought about the end of the alleged lost earthly continent of Mu in the Pacific Basin.[100] The annotations of the *Varo Edition* also observe that many ancient writings refer to an ancient "great war."[101] But as the terms "great bombardment" suggest, this war was fought by hurling projectiles at targets, and the following short quotation from one of the annotations suggests something else (and

---

[98] Joseph P. Farrell, *Secrets of the Unified Field: The Philadelphia Experiment, The Nazi Bell, and the Discarded Theory* (Kempton, IL: Adventures Unlimited Press, 2008), pp. 292-294.

[99] Jessup and The Anomalies Network, (Http://www.anomalies.net), *The Case for the UFO: Unidentified Flying Objects, the Varo Edition* (No date, ISBN 9781479131431), pp. 31, 35, 57, 59.

[100] Jessup, *The Case for the UFO: The Varo Edition*, pp. 71, 80-81.

[101] IBD, pp. 103-105, 115.

note, I am preserving, as did the Varo edition itself, the peculiar spellings, capitalizations, and punctuation of the annotations):

> *Balls* were compressed Earth, Used as Ammo for Force-"Guns" During "The Great War" The success of *that*, was so fast, that these were never used.[102]

In other words, some type of weapon utilizing a force field was propelling projectiles as kinetic impact weapons to their targets in a "rods of God" scenario. It is to be noted, once again, that longitudinal or gravitational waves would be capable of doing this.

The "rods of God" scenario is repeated in a rather lengthy quotation toward the end of the *Varo Edition* which makes it clear that the "balls of earth" were very substantial affairs (and again, I am preserving exactly the peculiar spelling, capitalization, and punctuation of the *Varo Edition* annotation):

> If the history of the Great War of the ancients were ever recorded, except by the black-tongued ones own tales. It would cause Man to stand in awe (or disbelieve) that such Huge Satelitic Masses were ever deliberately tossed throo this atmosphere in an attempt to Demolish all of the "Little Men" Great Works.[103]

The point of this little excursion into Ufology is simply this: that even here, in an unexpected context, one has both (1) references to an ancient war with weapons that by their very nature (2) required the use of longitudinal waves, or if one prefer, "electro-acoustic" or "electro-gravitic" waves in the medium in order to work.

(3) A Tangent into Science Fiction: Eando Binder's Strange Science Fiction Novel, "The Puzzle of the Space Pyramids"

While we are making unnecessary but nevertheless worthwhile trips into tangential fields, yet another and eerily uncanny reference to pyramids as "longitudinal" or "gravity" wave

---

[102] M.K. Jessup, *The Case for the UFO: The Varo Edition*, p. 70.
[103] M.K. Jessup, *The Case for the UFO: The Varo Edition*, p. 164.

manipulators was made in the science fiction novel by Eando Binder titled *The Puzzle of the Space Pyramids*. Briefly put, the plot surrounds the fact that as humanity launches manned space missions to the various nearby planets of the solar system, it repeatedly discovers pyramids on the various planets. As more is learned about them, a conclusion is eventually reached about their purpose and function. Here it is best to site the novel directly and at length in order to illustrate both its perspicacity as regards the weapon hypothesis, but its connection of the hypothesis to the destruction of a planet in our solar system:

> "But one clue was in each pyramid, on each planet. A set of figures. Mathematics is a universal language. These figures told how much *power* each apex-machine produced."
>
> "Power to do what?" we asked patiently.
>
> "To move a planet."
>
> "What kind of power is that?" we gasped.
>
> "Gravity-power," Halloway said. "This Jupiter pyramid was rated at three hundred twenty-five units of gravity-power."
>
> "Move a planet?" That suddenly soaked in, to Swinerton and myself. "*What* planet, for God's sake?"
>
> "Asteroidia," Halloway said, as casually as though telling us it was snowing outside. "The Planet that once existed between Mars and Jupiter."
>
> We had to pry the rest out of him. We were cruel about it, as poor Halloway was completely spent. He could hardly talk. But he gamely gave us the whole story. His eyes shone dimly, as though he had looked through some window into the hoary past. And we could see his brain was a little giddy, with things that sunned and were almost incredible.
>
> "The Martians achieved civilization and conquered space about seventy-five thousand years ago, in Earth's time-scale. For twenty-five thousand years they colonized, sometimes ruthlessly.
>
> "For instance, they enslaved most of the Venusian race, which was why the modern natives wanted to kill off the first Earth Expedition, thinking them the returning Martian overlords of legend.
>
> "Also, on Earth, they killed off Neanderthal man, for some unknown reason, which neatly solves that anthropological mystery of our past. Father isn't sure, but they may also have warred on Atlantis, later, and may actually have caused that gigantic continent to sink, by super-forces.

"And don't think the Martians didn't have super-forces. For they moved, or tried to move, a planet.

"Fifty thousand years ago, it happened. The fifth planet, Asteroidia, had a very eccentric orbit. In fact, at one point, it met and crossed Mars' orbit. Some of the asteroids today still do that very same thing, causing the Martian craters.

They also cross Earth's and Venus's orbits.

"Eventually, through the ages, the two planets were coming closer and closer to meeting at that danger point. Several previous near-skimmings had raised enormous tides in the then-existing Martian oceans, destroying lives and cities. But worse, it was estimated that after several hundred years the two planets would collide head-on. Their orbits would intersect. Mars would be utterly destroyed.

"Scientists put their heads together. They must destroy the fifth planet, or move it. Martians did not want to migrate from their home planet forever. So the scientists devised a daring scheme.

"They built pyramids on Mercury, Venus, Earth, Mars, and Jupiter. They were simply foundations to hold their apparatus, and the pyramid-form is the most sturdy of any geometrical shape.

"*The machines were—well, gravity concentrators, we might say.* It's head and shoulders above anything we know.

"It's gravity control—the one thing, like radioactivity, that Earth science can't seem to do a thing with. I don't want even to guess at it, but *somehow these Martian scientists took some of the gravity of a planet, and projected it as a beam, to do with as they wanted.*

"The machines were needed on the four inner planets, in that they were small bodies with comparatively small gravitics. Only one set was needed for the outer planets—namely on Jupiter, with its tremendous storehouse of gravity. But the idea was to get at Asteroidia from both sides. Perhaps on each planet they build hundreds of pyramids and machines. Those we've found are the few that survived. Most of them crumbled away, in fifty thousand years.

"Anyway, the machines were completed and installed. And then the great tug-of-war began. They were trying to tug the errant planet out of its predestined orbit, into a new one that would no longer endanger Mars....

....

> "But something unexpected happened. Asteroidia finally fell apart under the terrific strain. Or rather, it exploded, becoming the pieces we know today as the Asteroids.[104]

When I first read Sitchin's interpretation of the *Lugal-e* with its compelling suggestion that the Great Pyramid was itself a weapon, and then went on to read the other texts presented here as a wider corroborative context for the weapon hypothesis, and for the type of "longitudinal wave-in-the-medium" and "space-warp" physics they suggested, I had no idea of the existence of Binder's book, nor what it said.

What his book suggests, however, is that someone had perhaps read the same texts, and had come to similar conclusions long before, and presented the weapon hypothesis in the guise of fiction.

As will be seen in this book subsequently, the relationship of gravity to the Great Pyramid was not confined merely to Binder's science fiction, nor to a "paleophysics" interpretation of ancient texts, but was rather a consuming interest and preoccupation with the first formulator of a mathematical theory of gravity, Sir Isaac Newton.

F. The Topoogical-Analogical Code of Ancient Texts

What emerges from this examination of ancient texts is a particular method of decoding them, which may be conveniently summarized by the following table:

Term in Text	***Contemporary or Scientific Meaning***
Ocean, sea, or abyss	Space, abyss of space, ocean of space, space-time, the physical vacuum and/or vacuum flux
Island	Any agglutination of mass, such as a particle or planet

[104] Eando Binder, *The Puzzle of the Space Pyramids* (New York: Modern Literary Editions Publishing Company, Curtis Books, 1971), pp. 198-200, emphasis added. It should be noted that Binder's book was first printed in 1937, under the name Otto O. Binder.

Mountain, mound	Any agglutination of mass, such as a particle or planet
Primeval waters (Tiamat, Hebrew *tehom*, &c)	Space, space-time, vacuum
Sound	Any wave of compression and rarefaction(i.e., longitudinal wave) of any frequency and amplitude in any medium
Light, Eye	Any electro-magnetic sin wave with amplitude, frequency, and phase in a medium
Ocean or Water Surface	Plane of the Ecliptic
Firmament, Heaven, Upper World	Above (north) of the plane of the ecliptic
Underworld, hell, lower world	Below (south) of the plane of the ecliptic
The "four winds" or "four corners" of the Earth	The cardinal compass points, coordinates, and at a deeper level, the geometry of a spherically circumscribed tetrahedron or pyramid, the Great Pyramid as a "squared circle" and "cubed sphere" analogue of the earth.

1. The Thesis Expanded: Longitudinal Waves and The Lattice of Space-Time: Warp Equals Weapon

As was seen above, the weapon of a "sound-eye", along with the other textual references to sound and noise in the war, is a profound clue to the nature of the physics behind "the Great Weapon" and why it was also known as the "Great Affliction." The term "sound-eye" recalls Nikola Tesla's longitudinal electric waves, waves which I have characterized as "electro-acoustic" or "electro-gravitic." We shall see in a moment why all these terms are more or less synonymous and interchangeable. But basically, the idea of the "sound-eye" in the Edfu Temple texts, along with the hints from other texts previously examined, is a profound hint and suggestion that longitudinal waves in the medium or lattice of the medium of space-time was the basic operative principle of the Great Pyramid, and what made it work.

Such waves are remarkably flexible in their applied uses, enabling any device using them with sufficient power to accomplish a variety of modalities. Since a longitudinal wave is a wave of compression and rarefaction in a medium, then one can:

1) Stress a nodal point or points of the medium lattice work *directly*, and inside a material object beyond its ability to damp, blowing it up, whether that object be a nucleus or a planet;
2) By broadcasting such a longitudinal wave in the medium at varying strengths, one could:
   a) broadcast wireless power over great distances to any point, as per the wireless power transmission of Nikola Tesla (as we shall see later);
   b) broadcast a longitudinal wave to move particular physical bodies in a particular way, such as an interplanetary ship, or "hurling asteroids" as in Carlos Allende's "Great Bombardment", since a longitudinal wave in the medium is, in effect, a *space-time warp*; thus,
   c) at lower power a longitudinal wave in the medium would be a perfect means of interplanetary communication via a kind of telegraphy, since such waves do not weaken over great distances, and their pulses can be detected; and,
   d) the Weapon Hypothesis and Warp-interplanetary capability *are one and the same, since they come from the same physics and would require similar mass-energy conversion metrics*, which would explain *why the ancient texts strongly indicate the wars were of an interplanetary nature and involved other peoples than simply from Earth.* As will be seen in a subsequent chapter, if the communications, movement, warp and weapon function are to work, one will likely need a mass-energy conversion metric of a very large scale, and will hence need a coupled oscillator of planetary

> scale.[105] As will be seen, the Great Pyramid, alone of all pyramidal structures, has sufficient dimensional analogues of local celestial space, including the Earth and Sun, to suggest that it is the trigger on a weapons system that includes the planet and Sun themselves, which constitute the "gun" itself. To put this point differently, take away the weapon function, and one takes away the warp and communications functions, and vice versa.

To put all these points somewhat differently, longitudinal waves in the medium are the mirror image of Einstein's idea from General Relativity that large masses such as a planet or star can "bend" (or warp or compress) the space (and hence, the lattice work) around them, for longitudinal waves of compression and rarefaction in the medium do *essentially the same thing: they create regions of compression similar to a mass, and as such, they can manipulate gravitational effects by dint of their "acoustic" character.*

One and the same type of physics, in other words, does all these things, but that does not mean that any device employing this physics is a weapon, rather, it means that *any weapon employing this physics is capable of other, benign, applications.* The weapon function represents, paradoxically, the most highly unified and the most flexible application of the physics.

A very rough analogy or two may assist in making this point clearer. Suppose, in the first instance, that Outer Slobbovia has constructed an isotope enrichment plant and begins to enrich uranium. Outer Slobbovia assures the world that the enrichment plant is purely for peaceful purposes, to bring the uranium stocks to enough purity that it can be used in their plans to expand nuclear power plants for electrical power generation. The enrichment plant

---

[105] It is to be noted that in the space-warp paper by Mexican physicist Miguel Alcubierre, the metric requires a mass-energy conversion of the approximate size of the planet Jupiter in order to create a space warp, or as I am calling them, "electro-acoustic" or "electro-gravitic" compressional wave. In the re-working of Alcubierre's mathematics, NASA scientist Dr.Harold "Sonny" White concluded that the metric and hence the mass-energy conversion requirement was much too large, and was in fact much smaller and within the realm of feasibility for humans.

is definitely *not* for the purpose of making an atom bomb. Matters are not helped, however, by the fact that Outer Slobbovia has opened a heavy water production plant. This, too, the country explains is simply for the purposes of producing a heavy water moderator for its reactors.

Unfortunately for Outer Slobbovia, the General Secretary of the World Oligarchs' Organization (W.O.O.), Waam Pau Bang, suspects that Outer Slobbovia is lying, and sends a team of experts to the country to determine that it is *not* making atom bombs. An inspection of the isotope enrichment plant indicates that the uranium is being enriched by centrifuges, and only to about 33% purity of the isotope U-238. The heavy water plant is concentrating on deuterium, and no other evidences are found. Since U-238 is not the fuel for a uranium-based a-bomb, and all the deuterium based heavy water is going directly into the reactor pool, the W.O.O. team breathes a sigh of relief, and produces a report stating it's highly unlikely that Outer Slobbovia's project is for a bomb.

If, on the other hand, the W.O.O. team had discovered that the method of isotope enrichment was laser enrichment, and that Outer Slobbovia was enriching U-235 to about 96% purity, and converting its stocks of heavy water deuterium to lithium-deuteride, then it would have been justified in concluding that Outer Slobbovia was lying, and that it was not only trying to produce an atom bomb, but a bomb that used "boosted fission" from the addition of the lithium-deuteride, or worse, that it was trying to produce a full hydrogen bomb from the lithium-deuteride, using the atom bomb as its "fuse." The great purity of the enriched isotope plus the ability to produce a boosted fission bomb from the lithium-deuteride might also convince Waam Pau Bang and the W.O.O.'s experts that Outer Slobbovia might even be intending to produce a great number of "mini-nukes."

Now let us extend the analogy. Outer Slobbovia just announced it has constructed a massive machine for the manipulation of weather, explaining it needed to do so to control the weather in order to damp the extremities of its seasons of terrible floods in the spring and endless droughts in the summer into a less violent pattern in order to increase its crop yield. Once again, Waam Pau Bang and the plutocrats of the World Oligarchs'

Organization are concerned, and demand to inspect the machine. They are shown the master control of the machine, a dial that says "Gentle Rain, Steady Rain, Gentle Wind, Partly Cloudy, Cloudy," and "Temperate Sunshine" and another dial that says "Target Region." After inspecting the innards of the machine, the team determines that the machine is entirely benign, because that's the only kind of weather it can produce.

But suppose the team had found that the first dial in the weather machine control room read "Earthquakes, Geomagnetic Storms, Tsunamis, Monsoons, Hurricanes" and "Tornadoes" in addition to "Gentle Rain, Steady Rain, Gentle Wind" and so on, and was capable of producing all of these things. In that instance, the W.O.O would be entirely justified in concluding that the machine, *despite the benign uses to which it* ***might*** *be put, was a weapon*, and few rational people would conclude otherwise. The ability to dial up a tornado in a target region—whatever other types of weather it may be capable of producing—makes it a weapon.

In other words, *if a weapons use was or is part of the original function of a technology, then normal human practice is to classify it as a weapon, in spite of the other benign uses to which the same technology might be put. If a weapons use was* ***not*** *part of the original function of a technology without significant modification of the original design of the technology, then normal human practice would not classify the original technology as a weapon.*

The significance of this principle vis-à-vis the texts examined in this chapter is thus evident, for in no case is the technology in question—the pyramid—ever suggested to have been subsequently modified in order to *become* a weapon. The only modification represented in the texts is a modification *to destroy any possibility of its* ***use*** *as such by removal of essential and original components.*

Does this mean that *all pyramids are weapons*? No it does not, for in the first instance, no other pyramidal structure has the degree of dimensional analogues to local space as does the Great Pyramid. The Great Pyramid appears as something *sui generis* as far as the weapon hypothesis is concerned. Other pyramids may manipulate similar forces *in kind*, but none of them are described in the texts as doing so to the *same degree.*

We have now examined some relevant texts that not only associate a tremendously powerful weapon of the gods to "mounds, mountains, and pyramids" but in one case, that of the *Edfu* temple texts, like Ninurta's "Victory Seal", to the Giza compound and pyramids, and specifically, to "the Great Pyramid." The texts are also clear, if read with physics eyeglasses and not the muddy spectacles of Egyptology, that there was a specific type of physics being utilized.

With this survey of texts now completed, it is time to examine the structure itself.

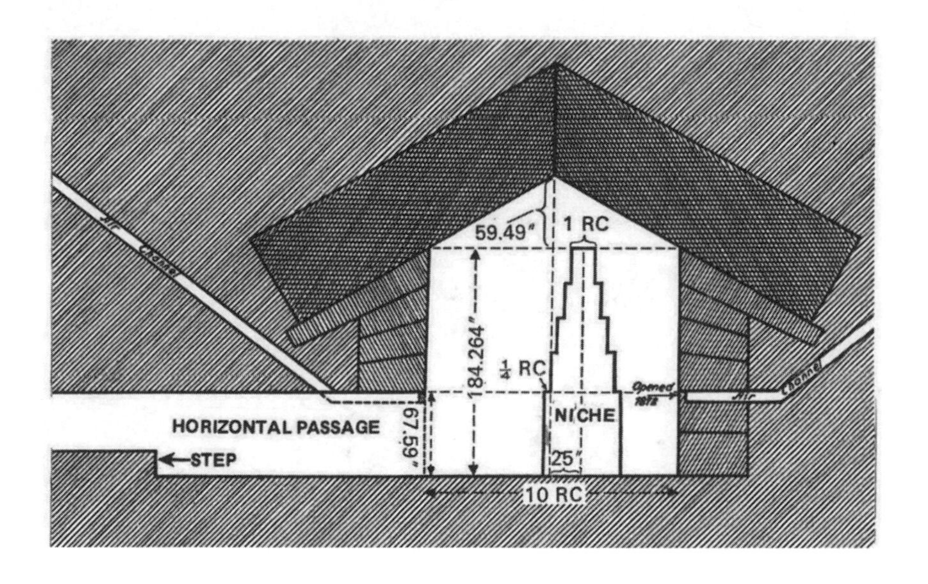

The "Queen's Chamber" and so-called "air shafts"

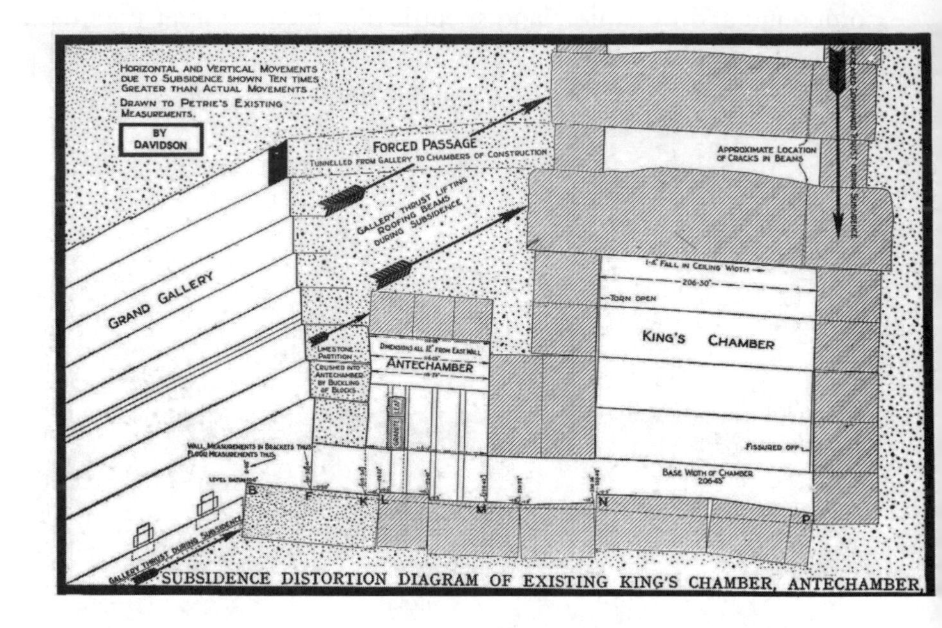

The Subsidence of the King's Chamber, possibly due to an internal explosion

4
CONCLUSIONS TO PART ONE

"... I wish only to demonstrate that, more than two thousand years ago, there existed a strong belief in a ***legacy*** *of ancient wisdom, which* ***pre-dated*** *Egyptian civilization. It is for this reason that we find such an amazing variety of ancient legends, from Greek, Roman and Arab times, which attempted in vain to identify the Great Pyramid's builder. It is apparent, therefore, that the idea of the Pyramids being the handiwork of a lost civilization is* ***a long-established belief****, not the inevtion of some 20th century treasure hunters."*
Alan Alford[1]

"A faint glimmer of light blooms in an antique papyrus authored by one Hardefef, a son of Cheops. He cited a very old writing that had been found in ***"the apex (benben) of the mountain (ben) uncovered by my father."*** *Breasted authenticates the story. The word* ***ben*** *or 'mountain' was quite commonly used to designate 'pyramid' by the ancient Egyptians.*
"This suggests that Cheops did not build the Great Pyramid, but uncovered it.... Anyway, Cheops had a name for refurbishing temples and other structures in and around Gizeh. There are Egyptian records of that. There is no record of his having built a pyramid." James Raymmond Wolfe[2]

WE ARE NOW IN A POSITION TO SUMMARIZE the basic direct and indirect textual and contextual evidence indicating not only that pyramids were understood to be weapons by many ancient texts, but that the Great Pyramid in particular was regarded as such.

1) Napoleon Bonaparte's expedition to Egypt not only resulted in the discovery of the Rosetta Stone which enabled the

[1]Alan Alford, *The Phoenix Solution*, p. 561, emphasis in the original.

[2] James Raymond Wolfe, "Experiments in Pyramid Power," in Martin Ebon, Ed., *Mysterious Pyramid Power* (New York: New American Library Signet Books, 1976, no ISBN), 73-90, p. 82, all emphases in the original.

subsequent decipherment by Champollion of Egyptian hieroglyphics for the first time in centuries, but it also resulted in artists' renditions and etchings of the Great Pyramid not only showing a clear line along the apothem, but also suggesting a very modern military appearance and "feel" to the structure to the extent that in the etchings the Great Pyramid strongly resembles modern American anti-ballistic missile phased array radar systems and installations.

2) The recent discovery by Scott Creighton of Colonel Howard Vyse's private, and hitherto unknown, private writings clearly demonstrates hesitation on Vyse's part over the correct rendition of King Khufu's (Cheop's) cartouche, and this hesitation all but proves that Vyse forged the inscriptions, as first suggested by Zechariah Sitchin and later more fully argued by Alan Alford.
3) With the collapse of Vyse's fraud, the one piece of "evidence" directly tying Cheops/Khufu to the Great Pyramid is removed, leaving questions of its dating and purpose open. Even radiocarbon dating dates the Pyramid to an earlier and pre-dynastic time by approximately four centuries, and that carbon dating may itself be subject to severe constraints on its accuracy if certain conditions were ever present within the structure.
4) Several ancient texts make it clear that pyramids were regarded as weapons:
    a) In Zechariah Sitchin's interpretation of the Mesopotamian text, the *Lugal-e*, the god Ninurta/Nergal clearly maintains that it is the "great Affliction" and "great weapon", and that an inventory was taken of its component parts which are various "stones"[3], with many of them being destroyed in order to render the structure inoperable.

[3] Q.v. Samuel Geller, "Die Sumerisch-Assyrische Serie *Lugal-E ud Me-lam-bi Nirgal*" in Bruno Meisner, *Altorientalische Texte und Untersuchungen* (London: Forgotten Books, 2018 ISBN 978-1-332-

b) Sitchin maintains that the text of the *Lugal-e* ***must*** refer to the Great Pyramid because Ninurta's "victory seal" reproduces the two large pyramids of Giza, which show the pyramids to have been smooth sided, rather than the "stepped" pyramidal form typical of Mesopotamian ziggurats.

c) Other texts, specifically, the *Edfu Temple Texts*, the *Enuma Elish*, *3rd Baruch,* the *Sibylline Oracles,* give significant clues as to the type of physics that was involved:

   i) The *Edfu Temple* texts clearly imply that mountains, mounds, or pyramids were identified by the Egyptians with the primeval creation itself, and with a tremendous war fought by means of that creative power with a weapon called the "Sound-Eye."

   ii) The term "Sound-Eye" suggests a kind of electro-magnetic ("eye) and acoustic ("sound") longitudinal pulse or wave was the principal means of the weapon's destructive power.

   iii) This understanding also dovetails with Sitchin's understanding of the *Lugal-E*, with its references to a "net" that was "grabbing and pulling" on Ninurta in order to kill him.

   iv) As noted in chapter three, longitudinal waves can also be used for long distance communication, and as a means of power and propulsion.

62308-2), pp. 277-316 for a complete phonetic transliteration of the text with German translation.

Intriguingly, Geller observes on p. 354 in his annotation to the text that there are *26* such stones that are "announced" by the text, which suggestively is one "stone" short of the 27 notches in the Grand Gallery. This may constitute another factor besides Ninurta's "victory seal" that suggested to Sitchin that the inventory was referring to the now-missing components of the Grand Gallery. As will be seen in part two of this book, I adhere to this theory, and view the names of these stones as the name of a *rank* of such components. On pp. 359-360 Geller provides a complete list of the "stones" of the *Lugal-E.*

v) The *Enuma Elish* also details the destruction of Tiamat, the exploded planet, by means of longitudinal waves in the medium that affected the very "bowels" or core of Tiamat, which began to experience earthquakes and massive "cries" or noise and sound, with Marduk driving one final pulse from his thunderbolt to destroy the planet;

vi) *3rd Baruch*'s Slavonic and Greek texts state that the builders of the Tower of Babel were trying to measure "the thickness heaven" and its composition by means of an auger, two references which suggest, again, a longitudinal wave, or warp in the medium, requiring a measurement or a "metric", and that this could only be achieved with an "auger" or *torsion.*

vii) Additionally, references such as "east" and "west" and "plains" or "thickness of heaven" and so on should also be taken or understood to be astronomical references, with "east" meaning "toward the Sun," and "west" meaning "away from the Sun" and "plain" meaning the "plane of the ecliptic", and :thickness of heaven" being a mass-energy conversion metric.

viii) Finally, it was shown that longitudinal waves in the medium is synonymous and identical with other expressions, such as "gravity waves," "Space warps," "electro-acoustic" waves or "electro-gravitic" waves. All these terms refer to longitudinal waves in a medium, and for our purposes the medium is space-time itself. This means that *the ability and power to create a space warp is at one and the same time the ability and power for faster than light travel* ***and*** *for mass planetary destruction.*

We are now in a position to ask the questions that dominate the next part of the book: Could the Great Pyramid have generated such waves? How might it have done so? And could it do so to with sufficient power to warp space-time itself and therefore to blow up a planet, or to wreak massive devastation on this planet?

All of the known indicators, as we shall see, are very suggestive of a "yes" answer to each of those questions…

Khufu/Cheops, Kahfre/Cephren, and Menkaure/Mycerinos

Nekoma, North Dakota Anti-Ballistic Missile Phased Radar Array Site;
The actual phased radar array is the pyramidal structure to the upper left of center

PART TWO: THE STRUCTURE ITSELF

"At the turn of the century Tesla was in the process of devising a means of wireless power transmission. ***The transmission involved the generation of longitudinal ether waves....****"*

Eric Dollard, *Condensed Introduction to Tesla Transformers* (Eureka, California: Borderland Sciences, 1986), p. 1, emphasis added

"The Persians and the great mass of the Magians, deny the Deluge altogether.... Further, they relate, that the inhabitants of the west, when they were warned by their sages, constructed buildings of the kind of the two pyramids which have been built in Egypt.... People are of the opinion that the traces of the water of the Deluge and the effects of the waves are still visible on these two pyramids half-way up, above which the water did not rise."

The Arab Historian Al-Biruni, *The Chronology of Ancient Nations*, trans. C. Edward Sachau, (London, 1879), cited in A. Pochan, *The Mysteries of the Great Pyramids:"The Luminous Horizons of Khoufou"* (New York: Avon Books, 1978, ISBN 0380-00881-5), pp. 65-66.

Sir Isaac Newton, 1642-1727

5
Dimensional Analogues and Analysis

"When Greaves chose the length of 693 English feet as his measurement for the side length of 'the greatest Egyptian Pyramid', as Newton called it, he was highlighting other features of the pyramids' design; and a wealth of information about the ancient world; and about the foundational principles of modern physics and mathematics."
Antonia Clarkeson[1]

"In order to establish reliable data towards proofs of his theory of universal gravitation, Newton needed to know the mass of Earth. The basis for a geodetically coherent system of measures might therefore be found in the ancient measurement of the length of a geographic degree. The ancient system of measurement standards was rooted in an intense and precise understanding of the physical world. By gathering the various standards into a coherent congruity, Newton would be expecting to facilitate the continuing scientific relevance of those measurement standards."
Antonia Clarkeson[2]

A. Herodotus: Radishes, Onions, Garlic, and Gravity

HERODOTUS' NAME IS FAMILIAR TO MOST PEOPLE even though many may never have read any of the comments or works for which he is justifiably famous, and by some lights, justifiably infamous. He was the classical world's premier historian and recorder of events and lore, though many modern

---

[1] Antonia Clarkeson, *Explaining Newton's* Disseration *upon the Sacred Cubit of the Jews* (http://users.tpg.com.au/adsley 22/FHG/9.%20Newton's20Dissertation.pdf), p. 7. Ms. Clarkeson's study is both essential for an understanding of Greaves' and Newton's study of the Great Pyramid and ancient metrology, and is also a brilliantly argued paper.

[2] Ibid., p. 21.

historiographers question his accuracy and reliability, and not a few have done so because of his propensity for the occasional tongue-in-cheek remark and tendency to "pull his readers' legs."

These tendencies are particularly in evidence when one considers exactly what he had to say about the Great Pyramid:

> Now they told me, that in the reign of Rhampsinitus there was a perfect distribution of justice, and that all Egypt was in a high state of prosperity; but that after him Cheops, coming to reign over them, plunged into every kind of wickedness. For that, having shut up all the temples, he first of all forbade them to offer sacrifice, and afterwards he ordered all the Egyptians to work for himself; some, accordingly, were appointed to draw stones from the quarries in the Arabian mountains down to the Nile, others he ordered to receive the stones when transported in vessels across the river, and to drag them to the mountains called the Libyan. And they worked to the number of a hundred thousand men at a time, each party during three months. The time during which the people were thus harassed by toil lasted ten years on the road which they constructed, along which they drew the stones, *a work, in my opinion, not much less than the pyramid*: for its length is five stades, and its width ten orgyae, and its height, where it is the highest, eight orgyae; *and it is of polished stone, with figures carved on it*; on this road then ten years were expended, *and in forming the subterraneous apartment on the hill, on which the pyramids stand, which he had made as a burial vault for himself, in an island, formed by the drawing of a canal from the Nile*. Twenty years were spent in erecting the pyramid itself: of this, which is square, each face is eight plethora, and the height is the same; it is composed of polished stones, and joined with the greatest exactness; none of the stones are less than thirty feet.
>
> This pyramid was built thus; in the form of steps, which some call crossae, others homides. When they had first built it in this manner, they raised the remaining stones by machines made of short pieces of wood: having lifted them from the ground to the first range of steps, when the stone arrived there, it was put on another machine that stood ready on the first range: and from this it was drawn to the second range on another machine: for the machines were equal in number to the ranges of steps; or they

> removed the machine, which was only one, and portable, to each range in succession, whenever they wished to raise the stone higher; for I should relate it in both ways, as it is related. *The highest parts of it, therefore, were first finished, and afterwards, they completed the parts next following; but last of all they finished the parts on the ground, and that were lowest.* On the pyramid is shown an inscription, in Egyptian characters, how much was expended in radishes, onions, and garlic, for the workmen; which the interpreter, as I well remember, reading the inscription, told me amounted to one thousand six hundred talents of silver. And if this be really the case, how much more was probably expended in iron tools, in bread and in clothes for the labourers, since they occupied in building the work the time which I mentioned, and no short time, besides, as I think, in cutting and drawing the stones, and in forming the subterranean excavation.[3]

It is intriguing to obscrvc that not much has changed in Egyptology since Herodotus' day, because in his account, the Great Pyramid functions as a tomb (even though the tomb is actually underground and on an artificial island created by diversion of the Nile river), it was built by thousands of workers quarrying rocks in distant quarries (the Arabian mountains) and presumably floated in "vessels" and brought over roads, then lifted into position by ramps, or ropes and pullies and being lifted on a machine. Herodotus' remarks remind one of the much more satirical version of Peter Lemesurier:

> The logic of the thing seems to defy all analysis.
>
> And so the historians, ably led by the classical Herodotus, have had their field-day. As well they might, bearing in mind that, even to Herodotus, the Pyramid's construction was already as remote in time as Herodotus himself is to us. Knowing precisely nothing about the project's origin, they have naturally fallen back on a process of wild extrapolation from their only slightly less scratchy knowledge of later dynastic times. The

[3] Herodotus, *Euterpe* 124-125, cited in William Kingsland, *The Great Pyramid in Fact and Theory* (Literary Licensing, no date, ISBN 9781497971417), pp. 5, 7, emphasis added.

Egyptians, it has been established, were obsessed with death and immortality, with the embalming of the dead, with preparations for life in the nether-world. Therefore the Great Pyramid Project represents that same obsession magnified to the nth degree. And so the scene described for us is a kind of gothic melodrama unequalled in its sheer antediluvian lunacy. The megalomaniac pharaoh Cheops, brooding over the fate of his own eternal soul, decides to throw his kingdom's entire resources into a colossal real-estate project designed purely to humour his own necromantic illusions of immortality. To satisfy this man's mere superstitious whim, thousands of slaves toil day after day to drag gigantic blocks of masonry up mighty ramps with the aid of nothing better than primitive sleds, levers, ropes and rollers. Overseers drawn straight from the serried ranks of Hollywood extras bark crude orders, wave cruder charts. The whips crack, the ropes creak, the tortured workers groan. For a brief moment in time the seething masses of ignorant sweating humanity toil with crude tools under the ancient sunlight, and then are swallowed up once more in the primeval mists of antiquity....

And the result? The Great Pyramid – a building so perfect and yet so enormous that its construction would tax the skills and resources even of today's technology almost to the breaking-point....

The sober truth is, of course, that no historian has yet advanced any explanation of the Great Pyramid's construction that is at all convincing. Nobody alive today knows for certain how the Pyramid was erected, how long it was in the building, how its near-perfect alignments were achieved before the invention of the compass, or how its outer casing was jointed and polished with such unsurpassed accuracy. Nor have historians succeeded in producing any convincing theory as to why such an enormous undertaking, combined with such incredible accuracy, should have been deemed necessary for the construction of a mere tomb and funerary monument to a dead king who in any case apparently never occupied it.[4]

---

[4] Peter Lemesurier, *The Great Pyramid Decoded* (New York: Avon Books, 1977 ISBN 0380-4303407), pp. 5-6.

Even Herodotus observes that simply constructing the road took ten years, and the Pyramid itself another twenty years, and that its cost in radishes, onions, and garlic – a diet designed it would seem to give the laborers incredible indigestion, explosive flatulence, and unbearable halitosis – was a veritable fortune in silver, an expense which, he also notes, did not include such things as the necessary *tools*, and clothing.

Like Lemesurier, Herodotus seems to have viewed all of this ancient Egyptology as a being more than just a bit incredible, because he notes that the construction of the necessary road for the stones *alone* was a task that in his opinion was not much less intensive than the pyramid itself, a criticism that many modern alternative researchers have repeated when confronted with the "ramps, ropes, and pullies" theories of the Great Pyramid's construction.

Notwithstanding the incredible nature of the amount of money expended, the amount of laborers and man-hours required (not to mention the vast fortunes that were probably made off of radish, onion, and garlic futures on the commodities markets!), the most incredible thing Herodotus records – no doubt with tongue firmly planted in cheek – is that *the vast structure was constructed from the top down rather than the bottom up.*

Herodotus may be having his little joke here. Then again, he may not, and may have heard the extraordinary tales of ancient Egyptian priests levitating the stones with "sound" and simply "walking" them rather effortlessly into place.

Still, building it "top down"!?

Even here however, this gravity-defying exercise is not the only thing connecting the Great Pyramid to the mysterious force of gravity, and maybe whatever Herodotus was told was an allusion to other, deeper connections of the structure to gravity that we have already encountered in the texts examined in chapter three.

As we shall now discover, there are even more direct and highly suggestive connections between the structure, its measures, and gravitation, connections that come together in the Great Pyramid's most famous scientific examiner, who discovered in its measurements dimensions that allowed him to complete his theory of gravity:

Sir Isaac Newton.

B. Greaves, Gravity, and Newton: Measuring the Immeasurable Pyramid

[The history of scientific curiosity concerning the Great Pyramid began after the Renaissance, with the well-known interest of Sir Isaac Newton in the structure being the most obvious example. Others, however, have also contributed significantly to the enormous pile of odd mathematical and physical "coincidences" that the structure holds. John Greaves, a young mathematician and astronomer educated at Oxford University, set off in 1638 searching for data that might establish the exact dimensions of the Earth. This was no idle or impractical scientific pursuit. In the aftermath of the discovery of the New World and the flurry of imperialism on the part of the European powers, navigation – and exact and precise charts – became essential. It was, literally, and for the English especially, a matter of national security.

> A clue to a possible solution had been postulated by Giraloamo Cardano, an astonishing Milanese physician and mathematician of the early sixteenth century and a close friend of Leonardo da Vinci's, who maintained that a body of exact science must have preexisted the Greeks. Cardano suspected that a degree of meridian (far more exact than that of Eratosthenes, Ptolemy, or Al Mamun) must have been in existence hundreds if not thousands of years before the Alexandrians and that to find it one must search in Egypt. Pythagoras was said to have claimed that the measures of antiquity were derived from Egyptian standards, themselves coped from an invariable prototype taken from nature. It followed that the pyramids might have been built to record the dimensions of the earth (sic) and furnish an imperishable standard of linear measure.[5]

[5] Peter Tompkins, *Secrets of the Great Pyramid* (New York: Harper and Row, 1971), p. 22. Tompkin's two books on the Great Pyramid and

Note carefully what the Renaissance scholars and scientists were saying: the pyramids, and in particular the Great Pyramid, were constructed as *analogs* or "scaled down" versions of the Earth itself. They were thus constructed *in a ratio or harmonic relationship to the Earth.* Had these scientists been familiar with the term, they might even have gone on to say that the Great Pyramid was constructed as an analog *computer*....

Though almost everything Greaves came across in his study "was a puzzle to him" he nevertheless, as a good scientist, carefully collected data and published the results. And he also made one of the first discoveries that would later be corroborated by others who entered the structure, a discovery that strongly suggests a machine function was the original purpose of the structure, though Greaves did not, apparently, pursue it.

Entering the Pyramid and beginning his descent down the Descending Passage, he was assailed by a blizzard of bats "so ugly and so large, exceeding a foot in length" that he decided to scare the bats away by firing his pistol. To his surprise, "the explosions reverberated like cannon shots in the restricted passage of the Pyramid."[6]

[[[7] Greaves' data, when published, caught the attention of a physicist working on a new theory of the most fundamental force of nature. The force was gravity, and the physicist was Sir Isaac Newton. From Greaves' carefully compiled mathematical data of the Great Pyramid, Newton deduced that it had been constructed using two different basic units of measurement, one a "profane" cubit, and the other a "sacred" cubit. "From Greaves's and Burattini's measurements of the King's Chamber, Newton

---

the Mexican Pyramids are thorough one-volume introductions to the subject, and are essential books and resources in any library of pyramidology, if one can find them!

[6] Peter Tompkins, *Secrets of the Great Pyramid*, p. 25.

[7] The section which appears here in double brackets is a page that was omitted in the original printing of *The Giza Death Star Deployed*, but which was restored in subsequent printings. For those possessing only the first printing, this is the original missing material.

computed that a cubit of 20.63 British inches produced a room with an even length of cubits: 20 x 10."[8] The importance of Newton's discovery cannot be overstated, for it contains two implications:

- The British system of measurement appeared remarkably close to a very ancient standard of measure; but,
- If the ancient system were used, most dimensions of the Pyramid could be expressed as whole numbers.[9]

The last point assumes great significance with respect to the Pyramid's possible "analog computer" function as well as with respect to "dimensional analysis", a technique theoretical physicists use to check the mathematical modeling of their theories. This will become evident in our review of Soviet pyramid research in chapter seven, and with the topological theories of Bounias and Krasnoholovits. In any case, as was also noted in *The Giza Death Star*, it appears that some similar whole numbers existed as an approximation of Planck's constant and other 'Planck units' in ancient times.

Newton derived the longer, or 'sacred' cubit from a comment made in the Jewish historian Josephus regarding the circumference of the pillars in the Jerusalem Temple. Estimating this cubit to be somewhere between 24:8 to 25.02 British inches, he published these results in an extremely rare paper bearing the somewhat lengthy title *A Dissertation upon the Sacred Cubit of the Jews and the Cubits of several Nations: in which, from the Dimensions]] of the Greatest Pyramid, as taken by Mr. John Greaves, the ancient cubit of Memphis is determined.*[10]

But why should a scientist of Newton's stature have spent so much time searching for an ancient unit of measure in the

---

[8] Peter Tompkins, *Secrets of the Great Pyramid*, p. 31.

[9] Egypt, it should be noted, represented most of the fundamental constants as ratios or fractions, rather than as decimals. For example, the constant π was represented as the ratio 22:7.

[10] Peter Tompkins, *Secrets of the Great Pyramid*, p. 31.

dimensions of the Great Pyramid? The answer is simple, but breathtaking:

> Newton's preoccupation with establishing the cubit of the ancient Egyptians was no idle curiosity, nor just a desire to find a universal standard of measure; his general theory of gravitation, which he had not yet announced, was dependent on an accurate knowledge of the circumference of the earth. All he had to go on were the old figures of Eratosthenes and his followers, and on their figures his theory did not work out accurately.
>
> By establishing the cubit of the ancient Egyptians, Newton hoped to find the exact length of their stadium, reputed by classical authors to bear a relation to a geographical degree, and this he believed to be somehow enshrined in the proportions of the Great Pyramid.[11]

Though Greaves' measurements were ultimately inaccurate, Newton's extrapolations from them were not. His figure for the Egyptian "sacred cubit" was very nearly perfect… and consequently, so was his theory of gravity.

Let us pause to consider what Newton's – as well as other early modern scientists' – preoccupations with the structure really mean. First, they indicate that these scientists knew about, and *took seriously*, the notion that there was an ancient scientific tradition that could possibly reconstructed by careful attention to texts and ancient structures. Secondly, and much more importantly, it means that Newton's theory of gravitation itself not only emerges in the context of such "paleoscientific pursuits", but more specifically is directly and immediately associated with the Great Pyramid. It is the first known and documentable example of a connection *between the structure at Giza and the force of gravity.*

The discovery of the measure of π in the structure had to wait until the nineteenth century, and the careful calculations of an amateur mathematician and astronomer named John Taylor. Wondering why the Pyramid would have been constructed along the peculiar – and very steep – angle of 51° 51′ Taylor concluded

[11] Peter Tompkins, *Secrets of the Great Pyramid*, p. 31.

that the surface area of each face of the structure equaled the square of its height. Thus, if he divided the perimeter of its base by twice the height, the quotient was 3.144, which was remarkably close to π.[12] The height of the Pyramid in relation to its base perimeter appeared to be that of a radius of a circle to its circumference! The Pyramid was, in effect, a *squared circle and a cubed sphere.*][13]

C. Deep Space, and Deep Physics Analogues
1. Newton, Gravity, and the Pyramid as a Scale Earth and Scale Universe

Greaves' and Newton's preoccupations with ancient systems of measure was not merely "to show the essential integrity and harmony of standards of measurement across the ages and across the nations and peoples."[14] Newton sought these measures in the Great Pyramid because he adhered to yet another doctrine of the lore of the Pyramid, namely, that the Pyramid was not only a scale representation of the Earth, but that within its measures and dimensions could be found even further analogues that made it a scale representation of the entire universe.[15] Australian scholar Antonia Clarkeson puts this important point in the following terms:

> In fact, (Newton's) texts can be taken as arguing the case for the antiquity of English measures and their Mesopotamian and

---

[12] Peter Tompkins, *Secrets of the Great Pyramid*, .p. 70.

[13] Joseph P. Farrell, *The Giza Death Star Deployed: The Physics and Engineering of the Great Pyramid* (Kempton, Illinois: Adventures Unlimited Press, 2003 ISBN 978-1-931882-19-3), pp. 65-69.

[14] Antonia Clarkeson, "Explaining Newton's Dissertation upon the Sacred Cubit of the Jews and the Cubits of Several Nations," (http://users.tpg.com.au/adsley/FHG/9.%20Newton's%
20Dissertation.pdf), p. 13

[15] Antonia Clarkeson, "Explaining Newton's Dissertation upon the Sacred Cubit," pp. 15,

> Egyptian origin. The texts form part of Newton's continuing interest in ancient standards of weights and measures. In order to establish reliable data towards proofs of his theory of universal gravitation, Newton needed to know the mass of Earth. The basis for a geodetically coherent system of measures might therefore be found in the ancient measurement of the length of a geographic degree.
>
> The ancient system of measurement standards was rooted in an intense and precise understanding of the physical world. By gathering the various standards into a coherent congruity, Newton would be expecting to facilitate the continuing scientific relevance of those measurement standards.
>
> Newton's analysis is essentially a deconstruction of the Pyramid's design; in the expectation that its elements can be remembered; and in accordance with the ancient approach to built work; that it reflects the geodetic and cosmic reality of its location, presenting as a scale model of a wider universe of which it stands as the cosmic centre. As such Newton is cngaged in an architectural exercise.[16]

Again, Newton's preoccupation may seem irrelevant to the whole association of the Pyramid with his theory of gravity, until one recalls yet *another* significant point about the controversy surrounding him and his theory once it was published and known, for at that time Newton was involved in a well-known and very heated controversy with the French over whether or not the Earth was slightly flattened at the poles, with Newton maintaining that it was, and the French, conversely, maintaining that it was elongated slightly at the poles.

How does the Pyramid fit into this?

Brace yourself…

2. And Nested Harmonic Relationships: Longitudinal Waves, and a First Look at the Pyramid as a Wave Guide and Wave Mixer

---

[16] Antonia Clarkeson, "Explaining Newton's Dissertation upon the Sacred Cubit, pp. 21-22.

Newton discovered that certain measures and dimensions of the King's Chamber were scaled-down measures of the height of the Pyramid itself, which in turn were accurate polar meridian measures of the Earth, *if the poles were flattened.* "This cryptic data" from the King's Chamber "affirms that the Egyptians were correct about polar flattening and that the *datum* was incorporated in the Pyramid's height."[17]

It should come as no surprise that Newton noticed the first example of what we shall later discover is *a very common and repeated phenomenon in the Great Pyramid: the relationship or connection or "nesting" of one dimensional analogue of local space in one location of the Pyramid, with that same dimension, or a multiple or divisor thereof, in another location in the structure.*

As will be seen in a subsequent chapter, this phenomenon is also a profound clue as to the function of the Great Pyramid, and a particularly strong indicator and argument for the weapon function. For the moment let us anticipate, in a very high overview, the detailed argument to be made later, for it again stresses the relationship of the Great Pyramid to gravity itself.

We saw that the texts strongly suggest a physics based on *longitudinal waves in the medium*, or on what I have also called "electro-acoustic" or "electro-gravitic" waves. Acoustic waves are, of course, waves in a sound-carrying medium, and as such are compressional or longitudinal waves. In effect, they are analogous to gravity waves, as gravity waves are analogous to music; both are compressional waves in a medium. Thus, when we say that certain dimensions of the Great Pyramid that are present in one area of the structure are reproduced elsewhere in the structure in some multiple or divisor of that dimension, *we are really saying that some dimensions of the Pyramid are harmonics of other dimensions of the Pyramid.* This was one of the things that always struck me about the building: as a pipe organist, it struck me very early on that so much of it appeared to be a coupled harmonic

---

[17] Antonia Clarkeson, "Explaining Newton's Dissertation upon the Sacred Cubit, p. 35.

oscillator of some other part or area of the structure. Thus, if one had a length of "x" units present in one area of the structure, that length – like the length of an organ pipe – would represent a resonator of a certain frequency. Elsewhere one might encounter that same dimension as 2x or 3x and so on, indicating lower and lower "frequencies" of the resonator, or as x/2, x/3 and so on, indicating higher and higher frequencies. The whole structure appeared as multitudes of such resonators, all in harmonic resonance to each other. It began to appear that the Great Pyramid was designed as an oscillator of the entire harmonic series.[18]

When we examine the structure in more detail, this is a profound clue that the measures of the structure are not meant merely to "memorialize" certain measurements or dimensional analogues, but rather they are there to be *oscillated*, i.e., to drive multiple longitudinal waves inside the structure, because if there purpose or presence in the Great Pyramid were merely to archive a standard of measure or some dimensional analogue of space, then one or two examples would have sufficed. It is the *redundancy* of such analogues that is telling us that their presence is *functional and not archival.*

This means, minimally, that the building is *a wave guide and wave mixer of longitudinal waves*. Information presented in subsequent chapters will only serve to confirm this argument.

D. Solar, Deep Space, and Deep Physics Analogues

There is yet another strange connection between the Great Pyramid, gravity, and even with more *avante garde* theories of

[18] Q.v. Antonia Clarkeson, "Explaining Newton's Dissertation upon the Sacred Cubit," p. 5, and Joseph P. Farrell with Scott D. deHart, *Grid of the Gods: The Aftermath of the Cosmic War and the Physics of the Pyramid Peoples* (Kempton, Illinois: Adventures Unlimited Press, 2011, ISBN 978-1-935487-39-5), pp. 244-249 for an interesting link to music between the Great Pyramid, and pyramidal structures in general, and the "musical mountain" of ancients.

contemporary physics. For example, for many years, Mars and space anomalies author and researcher Richard C. Hoagland has been arguing that the geometry of a spherically circumscribed tetrahedron in a rotating spherical mass such as a planet is a key to a hyperdimensional physics:

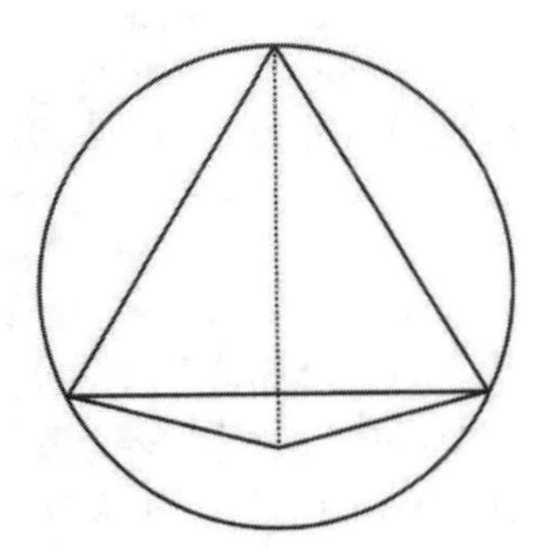

Spherically Circumscribed Tetrahedron

Noting that the vertices of such an inscribed tetrahedron occur at approximately 19.5 degrees north or south latitude (depending on the orientation of the vertex of the tetrahedron on the north or south pole of rotation of the body), Hoagland also notes that several planetary upwellings of energy occur, throughout the solar system, at this latitude, the Great Red Spot or storm on Juptier being the primary example of such energy at that approximate latitude. There are other signs of this geometry that we shall encounter later, but for our purposes, it is to be noted that the height of the King's Chamber was given by Greaves as "nineteen feet and a half". Greaves was not alone in noticing this tetrahedron-inside-a-sphere numerical constant, for Sir Flinders Petrie gave a height of 5,976.4124 millimeters, or using an English foot measurement of 306 millimeters, the height of the King's Chamber was a multiple of 19.53 of that measure.[19]

There is something else lurking in every dimension of the Great Pyramid involving π, and that is Plank's constant (ħ) and the

[19] Antonia Clarkeson, "Explaining Newton's Dissertation upon the Sacred Cubit,", p. 36.

constant 0.23706303 are always tied to π, for the following relationships are always present:

$$2 \times \hbar \times 0.23706303 = \pi, \text{ and}$$
$$2 \times \hbar \div \pi \times 0.23706303 = 1.$$[20]

We shall encounter many more such analogues in the Great Pyramid, but having established the link with gravity via Newton, and thereby with longitudinal wave forms via several ancient texts, we return, for a brief moment, to the strange comment of Herodotus that began this chapter.

E. Back to Herodotus' Strange Comment: Built from the Top Down and the Radiocarbon Dating of the Great Pyramids

We reproduce the relevant portion of Herodutus' remarks that is the focus of our attention here:

> This pyramid was built thus; in the form of steps, which some call crossae, others homides. When they had first built it in this manner, they raised the remaining stones by machines made o short pieces of wood: having lifted them from the ground to the first range of steps, when the stone arrived there, it was put on another machine that stood ready on the first range: and from this it was drawn to the second range on another machine: for the machines were equal in number to the ranges of steps; or they removed the machine, which was only one, and portable, to each range in succession, whenever they wished to raise the stone higher; for I should relate it in both wyas, as it is related. *The highest parts of it, therefore, were first finished, and afterwards, they completed the parts next following; but last of all they finished the parts on the ground, and that were lowest.* On the pyramid is shown an inscription, in Egyptian characters, how

[20] Antonia Clarkeson, "Explaining Newton's Dissertation on the Sacred Cubit", p. 62, taking the coefficient of ħ as 66260704. Clarkeson also observes that the Rydberg atomic constant is also present in the Great Pyramid(q.v. op. cit., p. 42)

> much was expended in radishes, onions, and garlic, for the workmen; which the interpreter, as I well remember, reading the inscription, told me amounted to one thousand six hundred talents of silver. And if this be really the case, how much more was probably expended in iron tools, in bread and in clothes for the labourers, since they occupied in building the work the time which I mentioned, and no short time, besides, as I think, in cutting and drawing the stones, and in forming the subterranean excavation.[21]

Perhaps Herodotus was simply "pulling our leg" that the outer casing of the Great Pyramid was built "top down," or perhaps he knew that the stones toward the top were geologically older, and thus from a deeper part, of the stone quarry than the ones at the bottom which were quarried first.

Or perhaps, without knowing he was doing so, he was preserving a rather different tradition, one that had, puzzlingly, maintained that the mortar between the stones of the building was older at the top than at the bottom, something that would throw off any radiocarbon dating by several hundreds of years and, indeed, indicate an older age at the top than at the bottom*, if there were a radiation source inside the Pyramid at any point in its past in the internal chambers, which are closer to the ground than the top.*

Radiation sources inside the Pyramid, longitudinal waves bouncing around and mixing inside it, and dimensional analogs of local geodetic and celestial space…

… all to bury a King?

Nonsense, says Christopher Dunn. It wasn't a tomb, it was a *machine…*

---

[21] Herodotus, *Euterpe* 124-125, cited in William Kingsland, *The Great Pyramid in Fact and Theory* (Literary Licensing, no date, ISBN 9781497971417), pp. 5, 7, emphasis added.

6
THE POWERPLANT OR MACHINE HYPOTHESIS

"A credible theory would have to explain..."
Christopher Dunn[1]

"In this system that I have invented it is necessary for the machine to get a grip of the earth, otherwise it cannot shake the earth. It has to have a grip on the earth so that the whole of this globe can quiver."
Nikola Tesla[2]

A. A Credible Theory

[AUTHOR AND ENGINEER CHRISTOPHER DUNN PROPOSES the most lucid account of the machine hypothesis in his book *The Giza Power Plant.* This crucial book in the growing literature on the Great Pyramid can only be summarized here, but every effort will be made to cite Dunn's actual words. It cannot be too strongly emphasized that this is only a summary of his work, and no summary can substitute for a careful study of his illuminating and important analysis. **(It is safe to say that Dunn's brilliant work is so magisterial that the nature of Pyramid research is forever changed, and that it is no longer possible for any rational and reasonable person to ignore the machine hypothesis.)**

---

[1] Christopher Dunn, *The Giza Powerplant: Technologies of Ancient Egypt* (Santa Fe, New Mexico: Bear and Company, 2003, ISBN 978-1-879181-50-9), p. 46.

[2] Nikola Tesla, trial transcript, in answer to a question from the bench; New York State Supreme Court, Appellate Division, Second Department: Colver Boldt Miles and George C. Boldt, Jr. as Executors of the Last Will and Testament of George C. Boldt, Deceased, Plantiffs_Respondents, versus Nikola Tesla, Thomas G, Shearman, et al. as Defendants-Appellants, 521-537, at line 529, cited in David Hatcher Childress, ed., *The Gantastic Inventions of Nikola Tesla* (Kempton, Illinois: Adventures Unlimited Press), p. 177.

For Dunn, any credible theory about the Great Pyramid would have to account for the following anomalous facts:

- The selection of granite as the building material for the King's Chamber. It is evident that in choosing granite, the builders took upon themselves an extremely difficult task.
- The presence of four superfluous chambers above the King's Chamber.
- The characteristics of the giant granite monoliths that were used to separate these so-called "construction chambers."[3]
- The presence of exuviae, or the cast-off shells of insects, that coated the chamber above the King's Chamber, turning those who entered black.
- The violent disturbance in the King's Chamber that expanded ts walls and cracked the beams in its ceiling but left the rest of the Great Pyramid seemingly undisturbed.
- The fact that the guardians were able to detect the disturbance inside the King's Chamber, when there was little or no exterior evidence of it.
- The reason the guardians thought it necessary to smear the cracks in the ceiling of the King's Chamber with cement.
- The fact that two shafts connect the King's chamber to the outside.
- The design logic for these two shafts—their function, dimensions, features, and so forth.

Any theory offered for serious consideration concerning the Great Pyramid also would have to provide logical reasons for all the anomalies we have already discussed and several we soon will examine, including:

- The antechamber
- The Grand Gallery, with its corbeled walls and steep incline.
- The Ascending Passage, with its enigmatic granite barriers.
- The Well Shaft down to the Subterranean Pit.
- The salt encrustations on the walls of the Queen's Chamber.

---

[3] The four chambers and granite monoliths that Dunn is referring to are the so-called "relieving" chambers above the Kng's Chamber.

- The shafts that originally were not fully connected to the Queen's Chamber.
- The copper fittings discovered by Rudolph Gantenbrink in 1993.
- The green stone ball, grapnel hook, and cedar-like wood found in the Queen's Chamber shafts.
- The plaster of paris that oozed out of the joints made inside the shafts.
- The repugnant odor that assailed early explorers.[4]

While Dunn's work admirably fulfills these criteria, there are nevertheless some omissions from this list that must be accounted for as well. What was all this engineering designed to do? Dunn's answer is obvious from the title of his work: it was to produce power. But power for *what?*][5] Power for whom? How *much* power?

More importantly, what is the purpose of all the nested harmonic relationships between parts of the building? Why are dimensional analogues of various parameters of local space-time present in the structure, parameters such as the distance to the Sun, the neutral gravity point of the Moon-Earth system, the velocity of light, the presence of analogues to other physical constants including those of quantum mechanics, and so on… why are these present in the structure at all? After all, what ordinary power plant includes the presence of such things in the production of power? As will be argued in the coming chapters, their mere presence in the structure means that this is no ordinary power plant, cogent and persuasive as Mr. Dunn's arguments are.

Additionally, the presence of dimensional analogues to local space, plus the clear and consistent association of pyramids with weaponry in ancient texts that we have already encountered, means that Mr. Dunn's argument will require significant modification.

However, because [so much of the weapons hypothesis is related to Dunn's work, the basic functions of the physical features of the Great Pyramid in his model will be summarized here before

---

[4] Christopher Dunn, *The Giza Powerplant,* pp. 46-47.
[5] Joseph P. Farrell, *The Giza Death Star*, pp. 180-181.

exploring how they may or may not relate to the weapon hypothesis **(in the remaining chapters of this book).** Dunn's hypothesis will be summarized here in more or less the order he himself develops it.

B. Extremely Close Tolerances and Some Provocative Questions

Engineers who have studied the Great Pyramid consistently come away amazed, and utterly baffled by its unusually close tolerances. Indeed, notes, Dunn, it was this factor that compelled his interest in the structure.

> Here was a prehistoric monument that was constructed with such precision that you could not find a comparable modern building. More remarkable to me was that the builders evidently found it *necessary* to maintain a standard of precision that can be found today in machine shops, but certainly not on building sites.[6]

But why were such tolerances necessary for a building that was **(supposedly)** designed principally as a religious tomb and/or astronomical observatory? **(Once again, the same question persists, and neither standard Egyptology nor much of the alternative revisionist hypotheses have a very good answer to it: why was the presence of such dimensional analogues in the structure necessary at all? And as for the tolerances,)** why have such tolerances at all? And how where they achieved?[7] Dunn's answer does not require the reader to subscribe to the dubious notion that the whole structure was built to such tolerances to ensure the Pharaoh's immortality!

> I consider two possible alternative answers. First, the building was for some reason *required* to conform to precise specifications regarding its dimensions, geometric proportions, and its mass. As with a modern optician's product, any variation from these specifications *would severely diminish its primary*

---

[6] Christopher Dunn, *The Giza Powerplant*, p. 51.
[7] Christopher Dunn, *The Giza Powerplant*, p. 56.

> *function.* In order to comply with these specifications, therefore, greater care than usual was taken in manufacturing and constructing this edifice. Second, the builders of the Great Pyramid were highly evolved in their building skills and possessed greatly advanced instruments and tools. The accuracy of the pyramid was normal to them, and perhaps their tools were not capable of producing anything less than this superb accuracy, which has astounded many over the years. Consider, for example, that the modern machines that produce many of the components that support our civilization are so finely engineered that the most inferior piece they could turn out is more accurate than what was the norm for those produced one hundred years ago. In engineering, the state of the art inevitably moves forward.[8]

These two provocative observations require some comment, beginning with Dunn's idea that the extreme tolerances may just be the coincidental effect of a society possessed of a highly developed skill in engineering.

But that is the point: If such skills exceed the capacity of our own most advanced construction capabilities, then one is dealing with a civilization more advanced than our own. This point is also highly significant, because the most pervasive employment of such construction tolerances in our own society is often in connection with military projects or extremely advanced optics, often both.

The second point is this: if these extremely close tolerances were necessary to the proper functioning of the structure, then one is faced with something of an anomaly for which no contemporary analogy actually exists, and so one is left to made educated conjectures. Our own contemporary notions of power and energy would not seem to require such close tolerances **(not to mention so many dimensional analogues of local space)** for the construction of a mere power plant, *unless* the paleoancient notion of power and energy was fundamentally different than our own, based on a kind of unified field physics that was *practical and testable*, a goal that we have not yet achieved. Indeed, Dunn never

---

[8] Christopher Dnnn, *The Giza Powerplant*, p. 64, emphasis added.

satisfactorily explains the necessity for such close tolerances **(and multiple dimensional analogues)** in a mere "power plant", **though as we shall discover, he *does* have an insight into why they are present.)** However, this is not a defect to his work, as he is concerned not so much with conjecture, but merely the evidence that the Pyramid was a type of machine involving enormous power output.

C. Advanced Machining and Ultra-Sonic Drilling

One of the most provocative and most thorough considerations of the advanced technology used in the Great Pyramid is Dunn's discussion of the evidence of advanced machining in the building.[9] In this respect, his analysis of the Coffer in the King's Chamber is the most anomalous of all the evidence for a sophisticated technology exceeding our own contemporary abilities.

> Along with the evidence on the outside of the King's Chamber coffer, we find further evidence of the use of high-speed machine tools on the inside of the granite coffer. The methods that were evidently used by the pyramid builders to hollow out the inside of the granite coffers are similar to the methods that would be used to machine out the inside of components today. Tool marks on the coffer's inside indicate that when the granite was hollowed out, workers made preliminary roughing cuts by drilling holes into the granite around the area that was to be removed.[10]

Dunn then reproduces the following figure **(overview of the coffer looking down from above on its interior):**

[9] Christopher Dunn, *The Giza Powerplant*, pp. 67-91.
[10] Christopher Dunn, *The Giza Powerplant*, pp. 79-80.

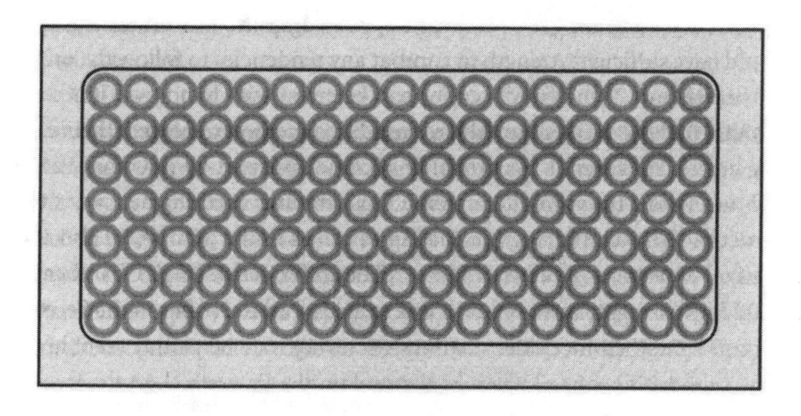

Dunn's Top-Down View of Inside of Coffer, Showing Tubular Drilling Marks[11]

The fact that the inside of the Coffer appears to have been drilled was one of the most anomalous facts noted by the famous nineteenth century pyramidologist, Sir William Flinders Petrie.

> To an engineer in the 1880s, what Petrie was looking at was an anomaly. The characteristics of the holes, the cores that came out of them, and the tool marks would be an impossibility according to any conventional theory of ancient Egyptian craftsmanship, even with the technology available in Petrie's day. Three distinct characteristics of the hole and core... make the artifacts extremely remarkable:
>
> - A taper on both the hole and the core.
> - A symmetrical helical groove following these tapers showing that the drill advanced into the granite at a feedrate of .10 inch per revolution of the drill.
> - The confounding fact that the spiral groove cut deeper through the quartz than through the softer feldspar.[12]

But in 100 years, as technology has advanced, the anomaly has deepened:

---

[11] Christopher Dunn, *The Giza Powerplant*, p. 80.
[12] Christopher Dunn, *The Giza Powerplant*, p. 84.

> In conventional machining the reverse would be the case. In 1983 Donald Rahn of Rahn Granite Surface Place Co. told me that diamond drills, rotating at nine hundred revolutions per minute, penetrate granite at the rate of one inch in five minutes. In 1996, Eric Leither of Tru-Stone Corp. told me that these parameters have not changed since then. The feedrate of modern drills, therefore, calculates to be .0002 inch per revolution, indicating that the ancient Egyptians drilled into granite with a feedrate that was five hundred times greater or deeper per revolution of the drill than modern drills! The other characteristics of the artifacts also pose a problem for modern drills. Somehow the Egyptians made a tapered hole with a spiral groove that was cut deeper through the harder constituent of the granite. If conventional machining methods cannot answer just one of these questions, how do we answer all three?[13]

Dunn's answer explains the advanced drilling technique used to hollow out the Coffer, but in the process, only validates the existence of an extremely sophisticated technology **(and of a science of longitudinal waves equally sophisticated)** in paleoancient times:

> In contrast, ultrasonic drilling fully explains how the holes and cores found in the Valley Temple at Giza could have been cut, and it is capable of creating all the details that Petrie and I puzzled over. Unfortunately for Petrie, ultrasonic drilling was unknown at the time he made his studies, so it is not surprising that he could not find satisfactory answers to his queries. In my opinion, the applications of ultrasonic machining is the only method that completely satisfies logic, from a technical viewpoint.
>
> Ultrasonic machining is the oscillatory motion of a tool that chips away material, like a jackhammer chipping away at a piece of concrete pavement, except much faster and not as measurable in its reciprocation. The ultrasonic tool bit, vibrating at 19,000- to 25,000-cycles-per-second (Hertz), has found unique application in the precision machining of odd-shaped holes in hard, brittle material such as hardened steels, carbides, ceramics,

[13] Christopher Dunn, *The Giza Powerplant*, p. 84.

> and semiconductors. An abrasive slurry or paste is used to accelerate the cutting action.
>
> The most significant detail of the drilled holes and cores studied by Petrie was that the groove was cut deeper through the quartz than through the feldspar. Quartz crystals are employed in the production of ultrasonic sound and, conversely, are responsive to the influence of vibration in the ultrasonic ranges and can be induced to vibrate at high frequency. When machining granite using ultrasonics, the harder material (quartz) would not necessarily offer more resistance, as it would during conventional machining practices. An ultrasonically vibrating tool bit would find numerous sympathetic partners, while cutting through granite, embedded right in the granite itself. Instead of resisting the cutting action, the quartz would be induced to respond and vibrate in sympathy with the high-frequency waves and amplify the abrasive action as the tool cut through it.[14]

An amazing anomaly indeed, for if ultrasonic drilling is a machining technique found only at the end of the twentieth century, then this would seem to imply that the paleoancient Vey High Civilization achieved *at least* a similar level of technological and scientific sophistication to our own.

Finally, there are two other facts about the Coffer that must be mentioned. First, the Coffer is one solid block of granite that has been hollowed out, probably either with ultrasonic drilling or with some technique as yet still unknown to us. And this raises a question: if the Coffer was meant to be a sarcophagus, why was it necessary for its builders to go through the extra complication of machining it in this fashion? Why not build it out of several pieces, as the Egyptians were known to do in other instances?[15] Second, the Coffer is a precisely machined object, not showing the slightest imperfection, *which means that it was constructed inside the King's Chamber*. Its builders intended for it to be precise for some as yet unknown reason.

---

[14] Christopher Dunn, *The Giza Powerplant*, p. 87.

[15] Christopher Dunn, *The Giza Powerplant,* p. 95. **(Of course, this may also be viewed as an argument that the builders of the Great Pyramid were not Egyptians.)**

> They had gone to the trouble to take the unfinished product into the tunnel and finish it underground for a good reason. It is the logical thing to do if you require a high degree of precision in the piece that you (sic.) are working. To finish it with such precision at a site that maintained a different atmosphere and a different temperature, such as in the open under the hot sun, would mean that when it was finally installed in the cool, cavelike temperatures of the tunnel, the workpiece would lose precision. The solution then as now, of course, was to prepare precision objects in a location that had the same heat and humidity in which they were going to be housed.[16]][17]

Here is should be noted that the Great Pyramid's builders constructed the building in such a way that the King's Chamber maintains a more or less constant cool temperature, a temperature that is very close to the mean thermal gradient or average temperature of the planet Earth. With this, it is once again apparent that the structure is deliberately designed as an oscillator of the planet itself, a feature that Dunn will make abundantly clear as we proceed.

D. How it Worked

[Dunn's analysis of the chambers and passages of the Great Pyramid and their possible functions, **(as far as it goes)** simply cannot be bested. **(Though it has some shortcomings which will be explored in a subsequent chapter, nonetheless)** it is the most comprehensive survey of their potentialities based upon known science and currently existing technology, and therefore what follows is but a crude summary of Dunn's excellent work. It is crucial to the basis of our speculations **(and arguments in subsequent chapters**), so some detailed understanding of Dunn's hypothesis is essential.

---

[16] Christopher Dunn, *The Giza Powerplant*, p. 97.
[17] Joseph P. Farrell, *The Giza Death Star*, pp. 182-188.

1. Missing Components and Many Possible Solutions

In addition to the advanced machining that so mystified Petrie, Dunn points out that the design of the inner chambers and passageways of the Great Pyramid seem to connote some functional purpose having little to do with the death-resurrection-Osiris mythology of ancient Egypt. "I became convinced that I was looking at the prints for an extremely large machine, *except this machine had been relieved of its inner components for some inexplicable reason.*"[18] This remark is truly astonishing, for nowhere in Dunn's work is any reference made to the ancient texts cited by Zechariah Sitchin that indicate that components were indeed removed from the Great Pyramid—some to be forever destroyed **(and some, which could not *be* destroyed to be forever hidden)**—by the victors in the "Second Pyramid War." Indeed, nowhere does Dunn refer to Sitchin's work at all. His approach, as an engineer simply examining the evidence the Pyramid presents, is to extrapolate from that evidence and known engineering and scientific principles the possible function of the Pyramid. On that basis, he concluded, "something is missing," corroborating independently the **(interpretation of)** the ancient texts cited by Sitchin.

But what kind of machine? Dunn maintains an open mind: "In proposing my theory that the Great Pyramid is a power plant, I am not adamantly adhering to any one proposition. The possibilities may be numerous. However, the main facts are inescapable, for they were noted many years ago, and it would be impossible for an open-minded, logically thinking person to disregard them."[19] If Dunn does not go all the way to the weapon hypothesis, he *does* allude to a potentially destructive use of the technology behind the Great Pyramid if not of the structure itself.[20]

---

[18] Christopher Dunn, *The Giza Powerplant*, p. 122, emphasis added.
[19] Christopher Dunn, *The Giza Powerplant*, p. 123.
[20] Christopher Dunn, *The Giza Powerplant,* pp. 243-245.

2. Some Elementary Physics: Coupled Harmonic Oscillators and Damping

The principle of a coupled harmonic oscillator in resonance to some fundamental "can unleash an awesome and destructive power."[21] The earth, as any college general science textbook will explain, is both a source of tremendous mechanical energy as well as of electro-magnetic energy, witness the enormous power unleashed in an earthquake or a thunderstorm. Normally, mechanical and electromagnetic energy propagate in two kinds of waves, transverse "S" **(or sine)** waves and longitudinal "P" **(or pressure-compressional)** waves. "Primary or compressional waves (P waves) send particles oscillating back and forth in the same direction[22] as the waves are travelling. Secondary or transverse shear waves (S waves) oscillate perpendicular to their direction of travel. P waves always travel at higher velocity than S waves and are the first to be recorded by a seismograph."[23]

---

[21] Christopher Dunn, *The Giza Powerplant*, p. 136. It should go without saying that the basic principle of resonance, as stated here, implies a potential for weaponization.

[22] The term "direction" may be misleading to some. What Dunn means is the particles oscillate along the same axis.

[23] Christopher Dunn, *The Giza Powerplant*, p. 126. The "wave-particle" duality of current quantum mechanics is well known.... Perhaps the duality is best expressed as the transverse-longitudinal wave duality.This form of stating the duality would thus seem to issue in a paradox: a photon of light would arrive at an observer before its detection or measurement by that observer. Superluminal velocity of longitudinal wave forms might thus form a basis of or connection to non-locality and photon entanglement.

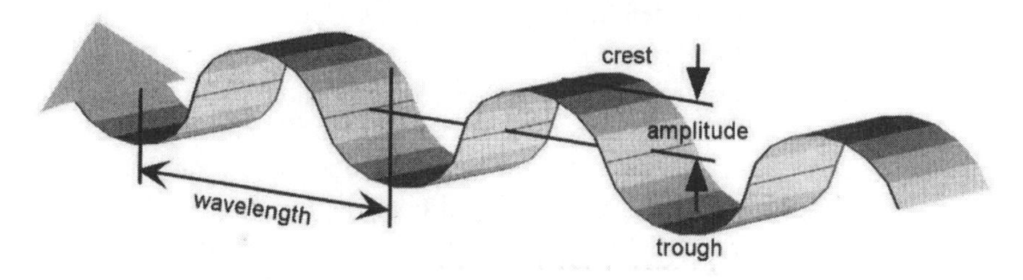

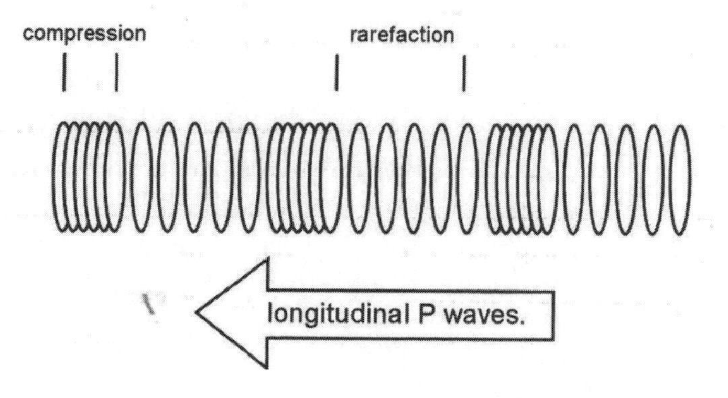

Dunn's Diagram of Transverse and Longitudinal Waves[24]

This relationship between mechanical, or acoustic types of waves, and electromagnetic waves, is deeply mysterious, but yet, is a commonplace that most people are familiar with. Dunn puts it this way:

> Turn on any motor or generator and you can hear the energy at work: the motor/generator will hum as it revolves. This hum is associated with the energy itself and not so much the movements of the rotor through the air. This phenomenon is evident when a motor stalls when the power is still turned on. When too great a load is put on a motor, and the motor stalls, the hum will become louder. The electrical and magnetic forces in the motor generate the sound waves. The earth itself, as a giant dynamo, produces similar sound waves.... Collectively known as an

[24] Christopher Dunn, *The Giza Powerplant*, p. 126.

electromagnetic 'cavity', the elements that make it up are the Earth, the ionosphere, the troposphere, and the magnetosphere. The fundamental frequency of the vibrations is calculated to be 7.83 hertz, with overlaying frequencies of 14, 20, 26, 32, 37 and 43 hertz.... The Earth's energy includes mechanical, thermal, electrical, magnetic, nuclear, and chemical action, each a source for sound. It would follow, therefore, that the energy at work in the Earth would generate sound waves that would be related to the particular vibration of the energy creating it and the material through which it passes.[25]

3. Piezo-Electric Effect

But why use granite, one of the most difficult materials to work with, in constructing **(portions of)** the Pyramid **(such as the King's Chamber)?** Very simple, says Dunn. Granite is composed of billions of tiny quartz crystals suspended in the surrounding rock. **(This fact, it should be noted, makes the entire edifice—for limestone also suspends quartz crystals—a non-linear material, and thus permits one to view the entire Pyramid as a giant crystal of acoustic non-linear acoustic metamaterial. The significance of this observation eludes most pyramid-as-machine interpreters, and its significance will be detailed in the next chapter.)** Thus, if one stresses the granite by pulsing it, each tiny quartz crystal would produce electrical output. This effect is known in physics as the piezoelectric effect.[26]

---

[25] Christopher Dunn, *The Giza Powerplant*, pp. 127-129.

[26] Piezo, meaning "stone". It is curious that the electrogravitics researcher and physicist Thomas Townsend Brown, whose other interests were known to have included UFOs, and who was elleged to have taken part in the design of the Philadelphia experiment, spent much of his last research investigating the electrical, magnetic, and acoustic properties of rocks, **(and this at approximately the same time the he was proposing his "Project Winterhaven" to the US Navy, to investigate his "gravitators"** ***and to investigate the use of longitudinal waves as a means of detection of atomic detonations and communications over inter-planetary and inter-stellar distances*****).**

> Any electrical simulation within the Earth of piezoelectrical materials—such as quartz—would generate sound waves above the range of human hearing. Materials undergoing stress within the earth can emit bursts of ultrasonic radiation. Materials undergoing plastic deformation emit a signal of lower amplitude than when the deformation is such as to produce cracks. Ball lightning has been speculated to be gas ionized by electricity from quartz-bearing rocks, such as granite, that is subject to stress.[27]

Dunn produces the following diagram to accompany this comment.

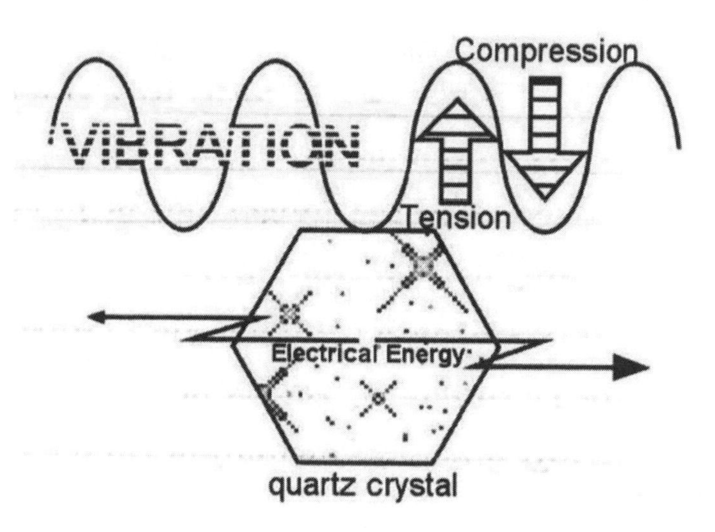

Dunn's Diagram of the Piezoelectric Effect[28]

So the choice of granite is now rather obvious, for the weight of the Pyramid Itself, pressing millions of tons of granite **(and limestone)** down through its stone courses, already places each tiny crystal under **(tremendous and)** constant stress. Add to this the fact that the Pyramid's ball-and-socket construction allows it to **(constantly quiver and)** *move* as a coupled harmonic oscillator means that all those quartz crystals **(embedded**

---

[27] Christopher Dunn, *The Giza Powerplant*, p. 129.
[28] Christopher Dunn, *The Giza Powerplant*, p. 129.

throughout the structure) are constantly being pulsed in resonance to the Schumann vibration of the Earth itself. But note also, that this stress may also generate a cloud of ball lightning, an ionized plasma of gas that itself contains energy. **(Dunn now approaches quite close to the weapon hypothesis with the following remarks, and these statements, it should be noted, are as close as Dunn comes to proposing a reason both for the nested harmonic relationships within the Pyramid, and for the dimensional analogues of geodetic and local celestial space:)**

> When we question *why* there is a correlation between the earth's dimensions and the Great Pyramid, we come up with three logical alternatives. One is that the builders wished to demonstrate their knowledge of the dimensions of the planet. They felt it necessary to encapsulate this knowledge in an indestructible structure so that future generations, thousands of years in the future, would know of their presence in the world and their knowledge of it.[29]
>
> *The second possible answer could be that the Earth affected the function of the Great Pyramid. By incorporating the same basic measurements in the pyramid that were found on the planet, the efficiency of the pyramid was improved and, in effect, it could be a harmonic integer of the planet.*
>
> A third alternative may involve both the first and second answers. The dimensions incorporates in the Great Pyramid may have been included to demonstrate the builders' knowledge or more importantly, to symbolize the relationship between the Great Pyramid's true purpose and the Earth itself.[30]

For reasons discussed in the **(next and remaining chapters of this book)**, I favor the second of these alternatives **(and in fact believe that any reasonable appraisal of either the machine or the weapon hypothesis would conclude that such dimensional**

---

[29] This is Dunn's statement of the time-capsule/observatory/ bureau-of-measures-and-standards hypothesis.

[30] Christopher Dunn, *The Giza Powerplant*, p. 134, emphasis has been added by me *in this book only*. No emphasis was added to this quotation in the original *The Giza Death Star*.

analogues are there solely and exclusively to improve the efficiency of the Pyramid's coupled oscillator function to "any possible target" in local space.)

4. The Grand Gallery:
An Acoustic Amplification Chamber and Helmholtz Resonators

Dunn is at his most brilliant when he analyzes the Grand Gallery and what its missing components may once have been. Noting that the ceiling tiles in the gallery tilt a an angle of approximately 45 degrees,[31] he observes that the Gallery is so constructed to be a massive acoustic amplification chamber, designed to amplify and reflect acoustic waves up the Gallery and toward the Antechamber.

> The mystery of the twenty-seven pairs of slots in the side ramps is logically explained if we theorize that each pair of slots contained a resonator assembly and the slots served to lock these assemblies into place. The original design of the resonators will always be open to question; however, if their function was to efficiently respond to the Earth's vibration, then we can surmise that they might be similar to a device we know of today that has a similar function—a Helmholtz resonator.[32]

A classic Helmholtz resonator is a hollow sphere, with an opening of 1/10th to 1/5th of the diameter of the sphere, usually made from metal but possibly from other materials.[33] Its size determines the frequency at which it resonates.

Dunn then builds his theory of what once existed inside the Grand Gallery.

> To extrapolate further we could say that each resonator assembly that was installed in the Grand Gallery was equipped with several Helmholtz-type resonators that were tuned to different harmonic frequencies. In a series of harmonic steps, each

---

[31] Christopher Dunn, *The Giza Powerplant*, p. 164.
[32] Christohpher Dunn, *The Giza Powerplant*, p. 165.
[33] Christopher Dunn, *The Giza Powerplant* p. 165.

> resonator in the series responded at a higher frequency than the previous one.... To increase the resonators' frequency, the ancient scientists would have made the dimensions smaller, and correspondingly reduced the distance between the two walls adjacent to each resonator. In fact, the walls of the Grand Gallery actually step inward seven times in their height and most probably the resonators' supports reached almost to the ceiling. At their base, the resonators were anchored in the ramp slots.[34]

He then produces the following diagrams of the resonator assemblies arrayed in the Grand Gallery:

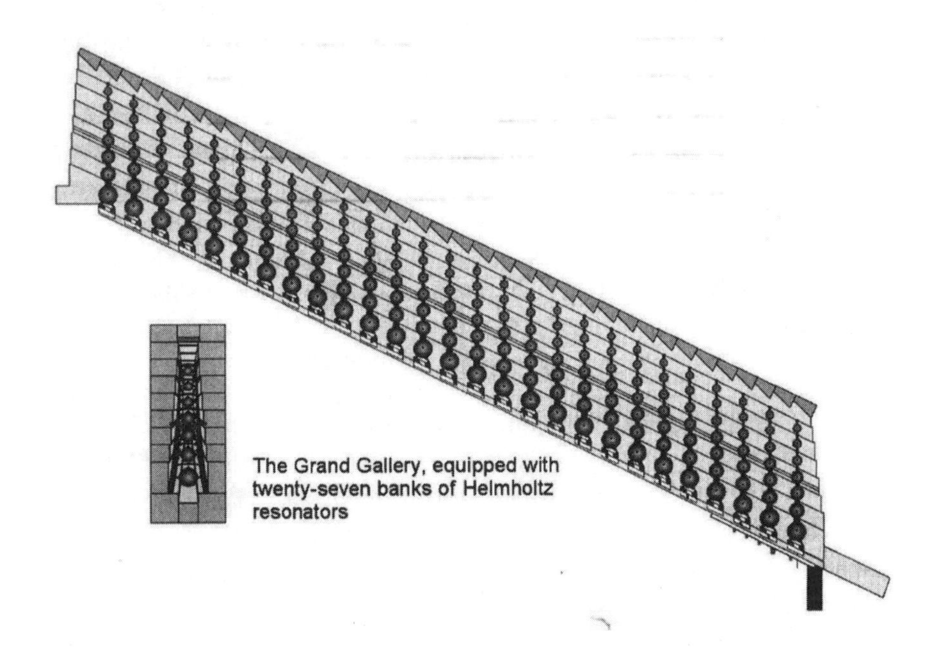

Dunn's Diagram of the Proposed 27 Helmholtz Resonator Assemblies in the Grand Gallery[35]

Let us pause at this juncture to observe some important points. First, note that Dunn has independently corroborated what Sitchin's **(interpretation of the *Lugal-E* indicated):** that certain

[34] Christopher Dunn, *The Giza Powerplant*, p. 166.
[35] Christopher Dunn, *The Giza Powerplant*, p. 168.

components, crucial to the functioning of the Pyramid, once existed inside the Grand Gallery itself. However, note also that there is a **(discrepancy)** between what Sitchin's **(interpretation of the *Lugal-E*)** states, and what Dunn hypothesizes once existed in the Grand Gallery. Sitchin's texts referred to the Gallery as having been bathed in multi-colored light from several "magic stones" or crystals arrayed in the notches on the side ramps. Dunn, conversely, and on the basis of sound engineering principles, concludes that the primary function of the Grand Gallery was as an acoustic harmonic amplification chamber. In **(the next chapter this book)** we will present a speculative resolution of this contradiction.

5. The Antechamber: Sound Baffle

With this very sound hypothesis in hand, Dunn next tackles the Antechamber. Drawing on Borchardt's hypothesis that the three slots did indeed once contain slabs that were lowered like a portcullis, he presents a credible theory of why such a machine would have been necessary. His solution is elegant. Whatever was raised or lowered in the slots in the antechamber functioned to block out sound waves coming from the Grand Gallery harmonic amplification chamber that were not of the desired **(frequencies).** By raising or lowering these objects, "sound waves with an incorrect frequency have wavelengths that do not coincide with the distance between the baffles and are filtered out."[36] Only the desired frequencies actually reach the King's Chamber.

6. The Air Shafts: Microwave Input and Output

When turning to the King's Chamber, there are three features that are the central focus of Dunn's attention. First, that the chamber itself resonates to the note f# on our own musical scale, a note that is a harmonic overtone of the Earth's own Schumann resonance. Second that the airshafts are not for the

---

[36] Christopher Dunn, *The Giza Powerplant*, p. 174.

purpose of admitting air at all. And finally, the coffer serves the purpose of coupling the input from the "air shafts" with the acoustical harmonic amplification coming from the Grand Gallery.

> In *this* power plant the vibrations from the earth cause oscillations of the granite within the King's Chamber, and this vibrating igneous, quartz-bearing rock influences the gaseous medium contained within the chamber. Currently this gaseous medium is air, but when this power plant operated, it was mostly likely **hydrogen** gas that filled the inner chambers of the Great Pyramid. The Queen's Chamber holds evidence that it was used to produce hydrogen... To maximize the output of the system, the atoms comprising this gaseous medium contained within the chamber should have a unique characteristic—the gas's natural frequency should resonate in harmony with the entire system.[37]

It takes litte imagination to understand the significance of using hydrogen as the gaseous medium inside the King's Chamber. Moreover, its presence there would explain the obvious melted look to the Coffer *if at some point an accident or deliberate explosive destruction occurred inside the chamber*.

As to how it all worked, Dunn puts it this way:

> Based on the previous evidence, sound must have been focused into the King's Chamber to force oscillations of the granite, creating in effect a vibrating mass of thousands of tons of granite. The frequencies inside this chamber, then, would rise above the low frequency of the Earth, through a scale of harmonic steps—to a level that would excite the hydrogen gas to higher energy levels. The King's Chamber is a technical wonder. It is where the Earth's mechanical energy was converted, or transduced, into usable power. It is a resonant cavity in which sound was focused. Sound roaring through the passageway at the resonant frequency of this chamber—of its harmonic—at sufficient amplitude would drive these granite beams to vibrate in resonance. Sound waves not of the correct frequency would be

---

[37] Christopher Dunn, *The Giza Powerplant*, p. 179, italicized emphasis Dunn's, boldface emphasis added.

filtered in the acoustic filter, more commonly known as the Antechamber.[38]

Thus the hydrogen gas would be both acoustically and electrically stressed or pulsed, **(making it in all likelihood a "cool" or endothermic plasma, and an electrically pinched one at that)**. As the hydrogen atoms absorbed this energy, their electrons would be pumped to a higher state, and undergo quantum jumps until they fell back to their ground state. As they did so, they would release a packet of energy in the microwave region of the electromagnetic spectrum.

7. The Coffer: the Optical Cavity of a Maser

Dunn notes that one of the peculiar features of the coffer, itself an object with so many dimensional ratios to the Earth, the solar system, and the galaxy, is also an optical cavity with concave surfaces at each end. Since the electrons of the hydrogen atoms can be stimulated to fall back to their ground state by an input signal of the same frequency, one has here all the makings of a maser: signal input, an optical cavity to cohere the emission of photons as electrons jump out of and back to their ground state, and, in the southern "air shaft" leading from the King's Chamber back to the face of the Pyramid, a horn antenna used to collect microwave beams. Thus the airshafts are not airshafts at all, but waveguides for microwave signal input and output. Dunn thus reasonably assumes, on sound scientific principles, that the Coffer was once correctly positioned exactly between the two shafts.[39] "The (originally smooth) surfaces on the outside of the Great Pyramid are 'dish-shaped' and may have been treated to serve as a collector of radio waves in the microwave region that are constantly bombarding the Earth *from the universe*. Amazingly, this waveguide leading to the chamber has dimensions that closely approximate the wavelength of microwave energy,

[38] Christopher Dunn, *The Giza Powerplant*, p. 183,

[39] Christopher Dunn, *The Giza Powerplant*, pp. 184-185.

1,420,405,751,786 hertz."[40] **(As we noted previously, however, this indentation along the apothem is true, *but only for the interior stone courses*. It does not appear to be true of the original casing stones. But in any case, Dunn's observation)** is tantamount to saying that the Pyramid's engineers built a structure that was designed to collect **(and possibly amplify)** the background radiation of the universe, radiation that most physicists currently believe was left over from the "Big Bang" itself, and that plasma cosmology maintains is the result of the electro-magnetic vorticular processes in evidence in galactic structures.

8. The Queen's Chamber: A Hydrogen Generator

On what basis does Dunn assume that hydrogen was indeed the gaseous medium inside the King's Chamber? Dunn is unhesitating in his belief that hydrogen was the gas used to power the Pyramid. "Without hydrogen this giant machine would not function."[41] Noting that early explorers to the Queen's Chamber beat a hasty retreat because of its overpoweringly unpleasant odor, Dunn speculates that a chemical reaction, such as between zinc and hydrochloric acid, were used to produce the hydrogen gas. Other chemical processes may have been used incorporating hydrogen sulfide, which would account for the odor.[42]

9. Meltdown, or Deliberate Destruction?

Having constructed this complex theory, Dunn then proceeds to account for the apparent violent disruption and dislocation evident in the King's Chamber: the slanted cracked walls, the melting of the Coffer, and the blackened limestone face on the surface of the **(upper)** interior of the Grand Gallery. These things he attributes to a "malfunction" that led to the hydrogen, "for some inexplicable reason," exploding in a ball of fire.[43]

---

[40] Christopher Dunn, *The Giza Powerplant*, p. 186, emphasis added.
[41] Christopher Dunn, *The Giza Powerplant*, p. 191.
[42] Christopher Dunn, *The Giza Powerplant*, pp. 200, 195.
[43] Christopher Dunn, *The Giza Powerplant*, p. 209.

Having undergone this accident, he theorizes that the Pyramid's builders then tunneled up to the Grand Gallery in order to make repairs, the occasion of cutting the much-debated "Well Shaft."

Here again, however, Dunn's theory contradicts the paleographic evidence marshaled by Sitchin. In the version preserved in **(his interpretation of the *Lugal-E*)** the Pyramid was entered for the deliberate purpose of inventorying its contents, and for earmarking some components for destruction and others for removal to be used in other devices elsewhere. In the face of clear evidence that there was some catastrophic destruction that took place on the inside of the Pyramid in the King's Chamber, and in the face of the paleographic testimony that this destruction was deliberate, I believe it is safe to say that the evidence of that destruction itself is the strongest corroboration of the textual evidence that states **(or suggests)** that its primary function was as a weapon.

10. The Other Pyramids, and Tesla

Dunn is alive to more sinister uses to his power plant theory, among them Tesla's use of pulsed harmonic vibrations.

> By applying Tesla's technology in the Great Pyramid, using alternating timed pulses at apex of the pyramid and in the Subterranean Chamber—a feature, by the way, that all the Egyptian pyramids have—we may be able to set into motion 5,273,834 tons of stone! If we have trouble getting the Great Pyramid going, there are three small pyramids nearby that we can start first to get things moving.[44]

These pyramids, he hypothesized, may have been employed "to assist the Great Pyramid in achieving resonance."[45]

But these insights, made almost in passing, raise as many questions as they answer, and in doing so, point out the relatively few weaknesses in Dunn's brilliant analysis.][46]

---

[44] Christopher Dunn, *The Giza Powerplant*, p. 149.
[45] Christopher Dunn, *The Giza Powerplant*, p. 219.

First, Dunn assumes that all the productions at Giza were Egyptian. But as I will argue subsequently, there are at least three layers of construction at Giza, and the Great Pyramid itself, and alone, constitutes the first of these. It is thus a self-standing machine, though I do not rule out machine purposes for the rest of the structures or the compound as a whole, as will be seen in the final chapter.

Secondly, and more importantly, Dunn does not really provide us with any reason or purpose for the power his machine is producing.

With this crucial and profound insight that Tesla's technology might provide an answer, Dunn leaves off, and does not pursue his insight any further.

When one *does* pursue it, one uncovers some astonishing correlations to the hypothesis that the structure was designed not only as a coupled harmonic oscillator of longitudinal "electro-acoustic" waves, one also is led very quickly to the weaponization of the phenomenon via Tesla's own words, and obtains insights into other possible functions of the structure.

To these we now turn.

---

[46] Joseph P. Farrell, *The Giza Death Star*, pp. 189-200.

7
THE WEAPON HYPOTHESIS

"Electrical capacity is to gravity, as inductance is to magnetism."
Michael Faraday

"'When the wave of expansion ebbs, suppose I explode another ton of dynamite, thus further increasing the wave of contraction. And, suppose the performance be repeated, time after time. Is there any doubt what would happen? There is no doubt in my mind. The earth would be split in two. For the first time in man's history, he has the knowledge with which he may interfere with cosmic processes!'
When Benson asked how long it might take him to split the Earth, he answered modestly, 'Months might be required, perhaps a year or two.' But in only a few weeks he said, he could set the Earth's crust into such a state of vibration that it would rise and fall hundreds of feet, throwing rivers out of their beds, wrecking buildings and practically destroying civilization."
Nikola Tesla[1]

"Commence primary ignition."
Faceless Voice, initiating activation of the Death Star, *Star Wars*

THE EPIGRAPHS ABOVE WHICH OPEN THIS CHAPTER make three things abundantly clear: (1) that a relationship between gravity and electromagnetism has been suspected for at least two centuries, (2) that Tesla thought it possible that one could actually bust the entire planet apart through mechanical resonance, and (3) that Hollywood, at least in the form of George Lucas' *Star Wars* and its "Death Star", was paying attention. When Tesla made his remarks to his laboratory assistant—remarks which subsequently found their way into the newspapers – Tesla was

[1] Margaret Cheney, *Nikola Tesla: Man out of Time* (New York: Bantam Doubleday Dell Publishing Group, Inc., 1981 ISBN 0-440-39077-X), p. 117.

constrained to admit to an anxious press and public that, to its great relief, the technology had not yet caught up to the principle.

Little did either Tesla himself or his anxious public realize, but within a few years, in his well-known Colorado Springs experiments, he would bring to fruition the first versions of a technology that could indeed split the Earth, a technology that has been massively misunderstood by people trying to view it as a conventional modification of a very conventional electrical technology. But it was neither a conventional electrical technology, nor even a conventional modification. There was nothing conventional about it, as engineer Eric Dollard and science writer Gerry Vassilatos demonstrated in their careful studies of Tesla's notes and apparatus.

A. Analyses of Tesla's Impulse Magnifying Transformer

When I decided to do this fourth book on the weapon hypothesis for the Great Pyramid, I debated whether to simply rewrite the sections concerning Telsa's Colorado Springs experiments and his electrical impulse technology from *The Giza Death Star* and *The Giza Death Star Deployed* into one coherent whole, or simply repeat them as they were originally written. Since nothing has really changed my view that Tesla's electrical impulse technology holds the key to the weapon hypothesis—after all, as we have seen, Christopher Dunn intuits the same thing—I have decided simply (1) to repeat what I originally wrote in *The Giza Death Star* about Gerry Vasilatos' presentation of Tesla's technology, (2) to repeat what I originally wrote in *The Giza Death Star Deployed* about Eric Dollard's review of that technology, (3) to incorporate the additional material from Tesla's 1915 trial transcript including his testimony about how his technology worked from my book *Babylon's Banksters*, before (4) including a new summary of all of this material here. In this way I hope my original reasons for focusing on Tesla's electrical impulse technology will become clearer, and in turn, that the weapon hypothesis itself comes into greater focus.

1. Gerry Vassilatos' Presentation of Tesla's Electrical Impulse Magnifying Transformer from the Original "Giza Death Star"

[The vast influence of the electrical work of Nikola Tesla is known to most informed people. Less well-known are the directions that Tesla's experimental research took from the close of the 19th century to the end of his life. Stories bordering on the mythological surround this period, stories of mysterious forces, and of government agents scurrying to confiscate his papers and notes upon his death. Whether those stories are true—and I am inclined to believe they are—the work of the last period of his life, incorporating some of Tesla's most brilliant experimental insights, affords a crucial look into the nature of physical reality as well as an understanding of the conflict between physics as an experimental science and as a theoretical and mathematical discipline. Moreover, Tesla's late work, and its subsequent profound misinterpretation by theoretical physics' orthodoxy, demonstrate the degree to which the received theories and paradigms of "normal science" can inhibit scientific insight, and be manipulated by vested power elites to close down lines of inquiry threatening to their own basis of power.

The author and science researcher Gerry Vassilatos has long investigated the "forgotten" highways of the physical sciences, and of the peculiarities of Tesla's last lines of research and his extraordinary claims with regard to its positive and negative potentialities. Vassilatos' work, *Secrets of Cold War Technology: Project HAARP and Beyond*, contains the clearest account in public literature of the "electrical impulse" investigations that so consumed Tesla in his later life. In this section we rely upon Vassilatos' account to summarize the experiment that led Tesla to investigate a whole new electromagnetic phenomenon, as well as the subsequent experiments he devised to confirm and expand his knowledge of it. These were precisely the experiments upon which Tesla based so many of his extravagant claims for a new source of limitless energy, as well as his seemingly fantastic claims for a weapon of mass destruction of planetary-busting power.

Vassilatos begins his account as follows:

> But while endeavoring toward his own means for identifying electrical waves, Tesla was blessed with an accidental observation which forever changed the course of his experimental investigations.... Part of this apparatus (was)... a very powerful capacitor bank. This capacitor "battery" was charged to very high voltages, and subsequently discharged through short copper bus-bars. The explosive bursts thus obtained produced several coincident phenomena which deeply impressed Tesla, far exceeding the power of any electrical display he had ever seen. These proved to hold an essential secret which he was determined to uncover.
>
> The abrupt sparks, which he terms 'disruptive discharges', were found capable of exploding wires into vapor. They propelled the very sharp shockwaves which struck him with great force across the whole front of his body. Of this surprising physical effect, Tesla was exceedingly intrigued. Rather like gunshots of extraordinary power than electrical sparks, Tesla was completely absorbed in this new study. Electrical impulses produced effects commonly associated only with lightning. The explosive effects reminded him of similar occurrences observed with high voltage DC generators. A familiar experience among workers and engineers, the simple closing of a switch on a high voltage dynamo often brought a stinging shock, the assumed result of residual static charging.[2]

This phenomenon led both power company engineers and Tesla to speculate on the reasons for this strange discharge. It should also be noted that the effect Tesla was obtaining bears some resemblance to the "electro-hydro-dynamic" phenomena being observed by **(Swedish plasma physicist)** Hannes Alfvén**.**

The theoretical and metaphorical framework in which Tesla framed his explanatory hypothesis for his next series of experiments points to a profound and persistent problem in theoretical physics, from relativity to quantum mechanics. At this

---

[2] Gerry Vassilatos, *Secrets of Cold War Technology: Project HAARP and Beyond* (Bayside, California: Borderland Sciences, 1996, 978-0-945685-20-3), p. 26.

juncture, it is important to recall that Tesla formulated his explanation *before* either of these theoretical bulwarks **(and in particular the concept of the velocity of light as an upper "speed limit")** were formulated.

> Tesla knew that *the strange supercharging effect was only observed at the very instant in which dynamos were applied to wire lines, just as in his explosive capacitor discharges.* Though the two instances were completely different, they both produced the very same effects. The instantaneous surge supplied by dynamos briefly appeared superconcentrated in long (power) lines. Tesla calculated that *this electrostatic concentration was several orders of magnitude greater than any voltage which the dynamo could supply. The actual supply was somehow being amplified or transformed.* But how?[3]
>
> The general consensus among engineers was that this was an electrostatic 'choking' effect.... Like slapping water with a rapid hand, the surface seemed solid. So also it was with the electrical force, charges meeting up against a seemingly solid wall. But the effect lasted only as long as the impact. Until current carriers had actually 'caught up' with the applied electrical field, the charges sprang from the line in all directions.... (Tesla) began wondering why *it was possible for electrostatic fields to move more quickly than the actual charges themselves, a perplexing mystery.*[4]

That is, Tesla knew that the electrical current moved at approximately the speed of light. But this in turn meant that the electrostatic field itself was moving at a superluminal velocity. **(Again, it is to be remembered that Tesla came to these conclusions *before* the Einsteinian-relativistic concept concerning the velocity of light had swept through theoretical physics.)**

---

[3] Tesla's mystification would be that of any competent physicist, for the increase of energy would appear **at first glance** to violate the Second Law of Thermodynamics.

[4] Vassilatos, *Secrets of Cold War Technology*, p. 27, all emphases added.

.... As Tesla saw it, the problem was that the brief, almost instantaneous application or "impulse" of electrical power impacting against the resistance barrier brought on an abnormally "electro-densified condition."[5] Through experimentation, Tesla determined "that he could literally shape the resultant discharge, *by modifying circuit parameters. Time, force, and resistance were (the) variables necessary to producing (sic) the phenomenon.*"[6]

(Swedish plasma physicist and Nobel laureate) Hannes Alfvén likewise maintained more or less the same thing: time, force, and resistance are variables that appear to follow laws that are *scale invariant from laboratory sized experiments to galactic superclusters.*[7] Note secondly that Tesla also makes the discovery that the geometric configuration of the circuit parameters is itself a factor determining how much energy was released in the discharge. **(To put this point differently, *the extra energy is coming not from the voltage input, but from the rest of the system, or circuit geometry, and more specifically, from the resistance itself, which in this case appears as an analogue of an inertial mass-energy conversion.*)**

Finally, an analogy may be helpful to explain the phenomenon. Tesla's reasoning behind the discharge was essentially this: At the very moment that the electrons of the current spark struck the wire or bus-bar, the geometry and density of the atoms in the bar effectively raised the resistance of the wire to infinity. No matter how high the current of the spark, the discharge still occurred. Electrons hit the resistance barrier, and splattered out from the surface of the bar in all directions perpendicularly **(to the velocity of impact, which again, was**

---

[5] Vassilatos, *Secrets of Cold War Technology*, p. 28.

[6] Vassilatos, *Secrets of Cold War Technology*, p. 28.

[7] This highly important point cannot be emphasized strongly enough, as this electrical and plasma approach *appears* to avoid the difficulties caused by the quantum mechanics-relativity split in contemporary theoretical physics by having an approach where the physics of very small systems and very large systems do not follow two different sets of laws, but one set that is scale invariant with respect to the size of the system under consideration.

near the velocity of light). We're all familiar with another form of this phenomenon. We've all climbed a diving board and jumped into a swimming pool, only to "belly-flop" and smack against the resistance of the surface of the water. At that very instant, no matter how fast we jump or how much we weigh, he hit up against that momentarily "infinite" inertial resistance of the water's surface, and send water splashing out in all directions around us. What Tesla did was to break the current at the very moment the electrons hit the surface of the wire, much like, if we could, at the very instant we hit the water, "run the film backwards" and do it over and over again in quick succession. With this simple analogy in mind, let us continue with Tesla.

In order to test the phenomenon further, Tesla resolved to repeat the experiments with direct current to eliminate the "backrush" to the dynamo caused by alternating current.

The result this time was even more astonishing:

> The sudden quick closure of the switch now brought a penetrating shock wave throughout the laboratory, one which could be felt both as a sharp pressure and a penetrating electrical irritation. A 'sting'. Face and hands were especially sensitive to the explosive shockwaves, which also produced a curious 'stinging' effect at close range. Tesla believed that material particles approaching the vapor state were literally thrust out of the wires in all directions.[8] In order to better study these effects, he poised himself behind a glass shield and resumed the study. Despite the shield, both shockwaves and stinging effects were felt by the now mystified Tesla. This anomaly provoked a curiosity of the very deepest kind, for such a thing was never before observed. More powerful and penetrating than the mere electrostatic charging of metals, this phenomenon literally

[8] In the original footnote here, I wrote "What Tesla meant by 'the vapor state' would be approximately what a quantum physicist would mean by quantum or sub-quantum particles, what we have called alternatively 'quantumstuff' or aether." I also think, however, that in its most basic sense Tesla is implying the results of longitudinal waves in the medium, or aether, itself.

> propelled high voltage out into the surrounding space where it was felt as a stinging sensation.[9]

In other words, throughout these experiments, Tesla was not only observing over-unity **(coefficients of performance)** energy output– getting more energy *out* than he was putting *in*—but feeling shockwaves *apparently oblivious to the normal shielding effects of matter.* He was getting more energy out of the system than he was putting into it, and feeling waves that traveled clean through solid objects like they were so much air. No wonder he was mystified! But he made the appropriate conclusion; *his system was not a closed system, but an open system, and he was somehow accessing a source of energy outside the system by dint of some inherent properties of the configuration of the system itself.*

In 1892, Tesla published a lecture in which he detailed these experiments. Titled "The Dissipation of Electricity", this lecture marks the point in Tesla's career where he abandoned research into high frequency alternating current for good in order to conduct a new line of experiments to describe the phenomenon of high energy direct current impulses and resulting "shockwaves."[10]

> He now prepared an extensive series of tests in order to determine the true cause and nature of these shocking air pulses. In his article, Tesla describes the shield-permeating shocks as 'soundwaves of electrified air.' Nevertheless, he makes a remarkable statement concerning the sound, heat, light, pressure, and shock which he sensed passing directly through copper plates. Collectively, they 'imply the presence of a medium of gaseous structure, that is, one consisting of independent carriers capable of free motion.' Since air was obviously not this 'medium', to what then was he referring? Further in the article he clearly states that 'besides the air, another medium is present.
>
> Through successive experimental arrangements, Tesla discovered several facts concerning the production of this effect. First, the cause was undoubtedly found in the abruptness of the

---

[9] Vassilatos, *Secrets of Cold War Technology*, p. 29.
[10] Vassilatos, *Secrets of Cold War Technology*, p. 31.

> charging. It was in the switch closure, the very instant of 'closure and break', which thrust the effect out into space. The effect was definitely related to time, *impulse* time. Second, Tesla found that it was imperative that the charging process occurred in a single impulse. No reversal of current was permissible, else the effect would not manifest. In this, Tesla made succinct remarks describing the role of capacity in the spark-radiative circuit. He found that the effect was powerfully strengthened by placing a capacitor between the disruptor and the dynamo. While providing a tremendous power to the effect, the dielectric of the capacitor also served to protect the dynamo windings.... The effect could also be greatly intensified to new and more powerful levels by raising the voltage, quickening the switch 'make-break' rate, and shortening the actual time of switch closure.[11]

It is now to be noted that the ... core of the Great Pyramid would function like a giant capacitor in Tesla's experiment, since its piezo-electric properties, under constant stress both from the mass of the structure as well as its resonance to the Schuman resonance would so stress the core as to build up **(a potentially)** phenomenal charge. *In other words,* ***(recalling the Edfu texts' "sound-eye", I am arguing)*** *that the Pyramid utilized some form of the same impulse energy discovered by Tesla,* ***(i.e., a longitudinal electrical wave, or "electro-acoustic" wave. It is to be noted that these are almost the same terms Tesla uses to describe the phenomenon he had observed.)***

Tesla's experiments also reveal more properties relevant to the Pyramid's function as a weapon. He discovered that it was possible to amplify the shockwave effect of the impulse by an *asymmetrical geometric arrangement of the system's components.*[12] "By placing the magnetic discharger closer to one or the other side of the discharging dynamo, either force positive or force negative vectors could be selected and projected."[13] That is, the power output of the system varied as a function of the

---

[11] Vassilatos, *Secrets of Cold War Technology*, pp. 31.32, emphasis in the original.

[12] Vassilatos, *Secrets of Cold War Technology*, p, 34.

[13] Vassilatos, *Secrets of Cold War Technology*, p. 34.

geometrical configuration of its components, exactly what our model of the hyperdimensional physics predicts. **(This observation, however, was made "hanging in the air" in the original *Giza Death Star*, and needs a bit more clarification here. Over the years I have remarked several times on the comment of the Hungarian electrical engineering genius, Gabriel Kron. Kron stated that *every* electrical system, no matter how simple and plain, was a "hyper-dimensional" machine *because* the mathematics necessary to describe electricity itself simply could not occur in just three dimensions. So much electrical engineering depends on the use of "imaginary numbers," and these, if looked at a certain way, are numbers that can only be conceived as inhabiting a fourth spatial dimension. We shall see later that there are other reasons for viewing the Pyramid as a hyper-dimensional machine. In any case,)** *precisely such asymmetry is found in the internal chambers of the Great Pyramid*, as well as the asymmetrical arrangement of the other structures at Giza. **(I do not believe that this arrangement is mere happenstance, but a deliberately designed feature of the structure for the above reasons, and for reasons that will be advanced later, but suffice it to note here that *some* of those reasons may involved an assymetrical structuring of the geometry of the system for the same or similar reasons as observed by Tesla.)**

More importantly, "Tesla found it impossible to measure a diminution in radiant force at several hundred yards. In comparison, he recalled that Hertz found it relatively easy to measure notable inverse square diminutions.... Tesla suspected that these effects were coherent, not subject to inverse laws other than those due to ray divergence."[14] We have already discovered that Dunn convincingly argues that a cohered microwave output **(a maser)** was involved in the Great Pyramid. But Tesla's experiments interest us for another reason, and that is the reliance of his open system upon geometric configuration to tap some unknown source of energy not subject to normal inverse square

[14] Vassilatos, *Secrets of Cold War Technology*, p.34.

relations. That energy was some sort of electro-acoustic *cohered* longitudinal wave. It remains to be shown how cohered microwaves and such "electrical impulses" might be connected and might have been weaponized in the Great Pyramid.

But Tesla's most astonishing hypothetical model concerns the actual over-unity energies he claimed to observe in his experiments.

> Actual calculation of these discharge ratios proved impossible. Implementing the standard magneto-inductive transformer rule, Tesla was unable to account for the enormous voltage multiplication effect. Conventional relationships failing, Tesla hypothesized that the effect was due entirely to radiant transformation rules, obviously requiring empirical demonstration. *Subsequent measurements of discharge lengths and helix attributes provided the necessary new mathematical relationship.*
>
> He had discovefed a new induction law, one whose radiant shockwaves actually *auto-intensified when encountering segmented objects. The segmentation was the key to intensifying the action. Radiant shockwaves* encountered an helix (sic) and *'flashed over' the outer skin, from end to end. This shockwave did not pass* ***through*** *the coil at all, treating the coil surface as an aerodynamic plane.* The shockwave pulse auto-intensified exactly as gas pressures continually increase through Venturi tubes. A consistent increase in electrical pressure was measured along the coil surface.... Tesla further discovered that the output voltages were mathematically related to *the resistance of turns in the helix. This resistance meant higher voltage maxima.*[15]][16]

If Tesla's electrical impulse magnifying technology is related to the Great Pyramid, it requires little imagination to understand *that the Great Pyramid is a segmented object par excellence, and that if segmentation actually amplifies the effect Tesla was observing,*

[15] Vassilatos, *Secrets of Cold War Technology*, pp. 36-37, All emphases added.

[16] Joseph P. Farrell, *The Giza Death Star*, pp. 255-263.

then the literally thousands of stones and segments comprising the Great Pyramid will amplify the effect to enormous proportions.

Vassilatos' survey of Tesla's impulse magnifying technology thus yields three conclusions, each of whose relevance to the weapon hypothesis of the Great Pyramid is immediately evident:

(1) The electro-acoustic shockwave phenomenon does not move *through* the circuit, but rather, over its surfaces;
(2) It does not obey standard inverse square law propagation; it is in fact a longitudinal wave, and as such, it is cohered and thus preserves its force over much greater distances;
(3) The force of the impulse can be modified by the geometric parameters of the circuit in two principle ways:
  (a) By asymmetrical arrangement of some of its components, as we argue that the internal chambers, and in particular the Grand Gallery, and King's Chamber, fulfill; and by,
  (b) The amount of segmented items the impulse encounters as it moves over the circuit, such as the number of windings in the coil helix, or in the Pyramid's case, the sheer thousands of stones or segments comprising the structure, and in its layered internal stone courses.

We will have occasion to return to these conclusions later in this chapter, but for the present, it is necessary to add more pertinent details from engineer Eric Dollard's study of Telsa's electrical impulse system.

2. Eric Dollard's Presentation of Tesla's Electrical Impulse Magnifying Transformer from the Original "Giza Death Star Deployed"

[Engineer Eric Dollard is unique in having successfully reproduced… Nikola Tesla's direct current electrostatic impulse technology and its anomalous results. His work will be extensively cited without commentary, with commentary following.

We begin, then, with Dollard's exposition of the Tesla Direct Current Impulse Magnifying Transformer:

> At the turn of the century Tesla was in the process of devising a means of wireless power transmission. The transmission involved the generation of longitudinal ether waves.... (1) *Tesla claims that the waves from his transformer propagate* ***at π/2 the velocity of light****. It is interesting to note that the velocity measure on the Tesla coil is also π/2 greater than the velocity of light but this does appear to be a phase velocity rather than a group velocity.*
>
> In his writings Tesla indicates some seemingly impossible phenomena surround the emanations from the spherical terminal capacitor, and I have determined this to be true by experiment. One is that the power gradient (Poynting vector) is in the same axis as the dielectric flux gradient. The other is the slow formation of a conductive area surrounding the sphere that is not ionic in nature (in other words is not a spark or glow discharge).
>
> Contrary to popular belief, the Tesla transformer is not a steady state device but a magnifier of transient phenomena. Also it does not behave like a ... (L.C. network) nor a transmission line, (2) *but more like a unique type of wave guide*. If all parts of the system are designed properly the (electromotive force) and hence (3) *dielectric flux jumps from zero to an enormous value almost instantaneously, thereby producing an almost inconceivable displacement current into space. The transformer is then basically a device for rapidly discharging the capacitor bank nearly instantly into free space, producing an enormous dielectric shock wave similar to a sonic boom.*
>
> Because the dissipation of the transformer is for all practical purposes negligible, the energy keeps increasing at a linear rate per cycle of oscillation, thereby accumulating a gigantic quantity of electrical energy.((4) *A form of laser action may be possible.*)
>
> (5) *In order for the transformer to resonate with the planet the energy storage in the active region that grows around the sphere terminal must equal the conjugate energy storage of the earth,* a stiff requirement.
>
> It is interesting to note that dielectric breakdown in this active region grow(sic.) into a log periodic form based on x^2-x=1 as the log base. (6) *This will be recognized as the transcendental PHI*

or Golden Ratio. (7) In glow discharges the ions of metallic elements form stable spheres of diameter inverse to the atomic weight of the element involved.

The transformer's principals (sic) of operation are as follows:

(8) *The first requirement is the sudden collapse of an energy field thereby producing a sudden impulse of energy,* second is the transforming properties of the odd harmonic order single wire delay line (coil) which allow for the production of enormous (electromotive force) and (magnetic motive force), and third, the dielectric phenomena surrounding the free space capacity terminal.[17]

The following implications emerge from this abstract:

(1) The superluminal velocity of the energy impulse **(i.e., π/2 the velocity of light, a value which has a distinctive relationship to the Great Pyramid, as we shall see later in this chapter)** was a conclusion that Tesla drew from observation and inference, since there does not yet exist a viable way to measure one-way propagation **(of the velocity)** of light and presumably of superluminal phenomena. Tesla's reasoning was that since the force propagated by the impulse did not diminish in accordance with the inverse square law, it may be inferred that it was a form of "time reversed" and hence superluminal wave, **(or perhaps better put, a phenomenon of phase conjugation).**

(2) It will be recalled from descriptions of this transformer given **(by Vassilatos in the previous section)** that the discharged impulse was not conducted *through* the coils, but "aerodynamically" *over the surface* of the coils. Hence, the coils may be viewed as types of "virtual crystals" or "electric lenses", or as Dollard puts it, "wave guides." This point becomes important in subsequent considerations.

---

[17] Eric Dollard, *Condensed Introduction to Tesla Transformers* (Eureka, California: Borderland Sciences, 1986), pp. 1-2,5, emphases added.

(3) Telsa's initial transformer was no taller than a child and produced massive shock waves. His subsequent famous experiments with a much larger version of this device in Colorado Springs produced lightning displays and shock waves visible and felt miles away. It stands to reason that if the Great Pyramid incorporated similar design features for a similar purpose, that its discharged impulse would be that much greater given its much more massive size than Tesla's Colorado Springs apparatus.

(4) While Dollard does not indicate the basis of his statement that "a form of laser action may be possible" it is relatively easy to infer why he might say this. The lack of inverse square energy dissipation effects in the impulse indicates a phenomenon exhibiting coherence properties. The harmonic nature of the transformer suggests that coherence of the phenomenon may be harmonically derived in some fashion.

(5) It is highly significant that the Tesla impulse magnifying transformer must be resonant to the energy of the Earth. The Great Pyramid is resonant to the mass and thermal gradients of the Earth and, as is well known, incorporates harmonics of various terrestrial and celestial geometries in various dimensional measures of the structure **(and is resonant to the Schumann cavity resonance of the planet).** *In short, the Tesla transformer and the Great Pyramid both appear to be electro-acoustic coupled oscillators to the Earth.*

(6) The log periodic growth of the energy buildup in the conductive "corona" around the transformer according to the ϕ ratio strongly suggests that the presence of the same number in the Great Pyramid has the *functional* purpose of achieving the same energy buildup, though of course on a much more massive scale **(by way of making the oscillator more efficient).**

(7) The inverse diameter relationship of the ionic sphere (of whatever element has been chosen as the dielectric) to the atomic weight of that dielectric element suggests a

relationship between the lattice structure of the elements themselves and the harmonics of the impulse waveform....

(8) The requirement of a very sudden field collapse in order to produce the longitudinal electric impulse wave suggests that a curious and surreal "darkening" would occur in the region of the device during discharge. This same phenomenon may be seen on films of nuclear explosions as the field collapses and space itself experiences a rarefaction and compression, and was attested by Tesla himself... Others working with different aspects of the impulse phenomenon also record similar darkening effects associated with field collapse **(as we shall shortly see)**.

Dollard then makes a significant statement concerning the field gradient of the electric potential of the device: "It would seem possible (for) the gradient to continue to increase beyond the dielectric terminal." In that instance the electro-motive force "also becomes *greater farther from the terminal*, possibly reaching astronomical proportions."[18] In other words, *great potential can be built up in a target region simply by the transmitter's distance from it.* **(This is a classic signature of a non-linear, and indeed phase conjugate, system, as)** potential becomes a function of distance from the dielectric terminal or antenna. Obviously, (**such a system**) makes it entirely feasible that it could have had interplanetary uses.

Considered as a system, the Tesla impulse magnifying transformer

> Can be divided into FIVE distinct components:
> 1) EARTH
> 2) REFLECTING CAPACITANCE
> 3) ENERGY TRANSFORMER
> 4) COUPLING TRANSFORMER

[18] Dollard, *Condensed Introduction to Tesla Transformers*, p. 10, emphasis added.

5) RESONANT COIL...[19]

Elsewhere Dollard is more specific:

> It is quite possible that the magnetic gradient and force will increase as the wave penetrates the earth. Hence the 5 sections of the Tesla transformer:
> 1. Earth
> 2. Primary system/power supply
> 3. Secondary wave coil
> 4. Tesla or magnification coil
> 5. Dielectric antenna.[20]

This catalogue of systems components now permits a schematic comparison of the Great Pyramid and a Tesla Magnifying Transformer, utilizing an adaptation of Dollard's schematic of the latter, and Dunn's schematic of the former **(see the next two pages).]**[21]

The significant differences between the two systems should also be noted in addition to the significant parallels and analogues, and chief among these is the fact that the Great Pyramid, at least in the present shell that sits at Giza, is constructed almost entirely from non-linear piezo-electric materials, i.e., a granite and limestone matrix in which are embedded *linear* materials in the form of tiny quartz crystals. Tesla's impulse magnifying transformer, on the other hand, is constructed almost entirely of *linear* materials, i.e., metals, with a regular lattice structure. It is as if some ancient engineer or engineers had taken the entirety of Tesla's impulse system, and magnified it yet again my making it an entirely non-linear system exponentially more powerful.

---

[19] Dollard, *Theory of Wireless Power* (Eureka, California: Borderland Sciences, 1986), p. 7.

[20] Dollard, *Condensed Introduction to Tesla Transformers,* p. 11.

[21] Joseph P. Farrell, *The Giza Death Star Deployed* (Kempton, Illinois: Adventures Unlimited Press, 2003, ISBN 978-1-931882-19-3), pp. 197-200.

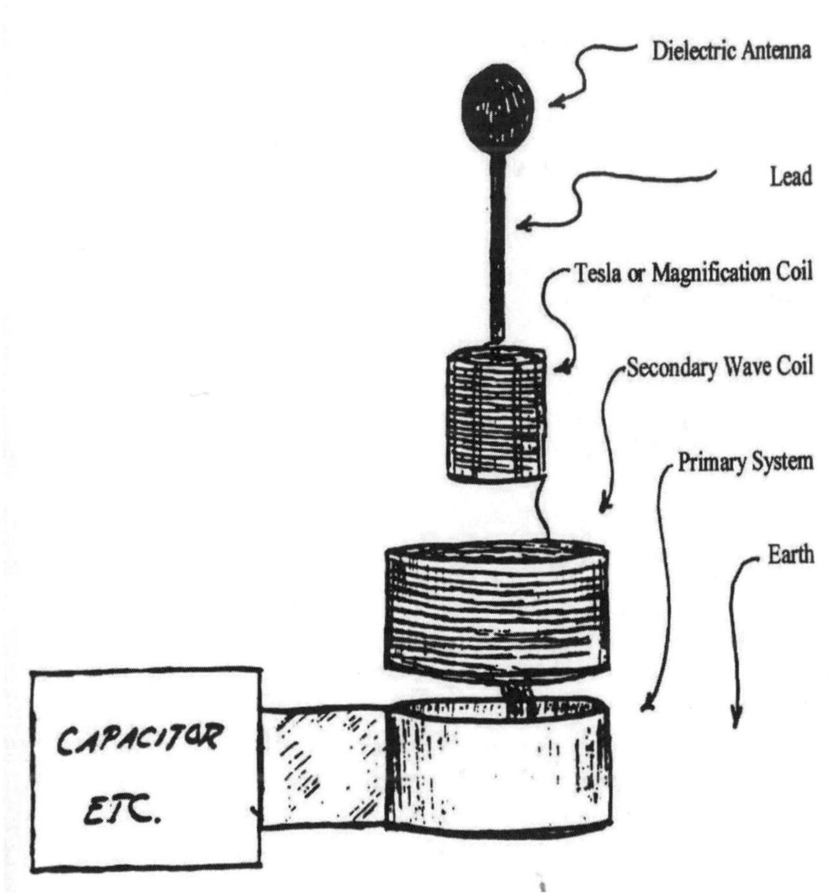

Eric Dollard's Schematic Diagram of the Components of a Tesla Impulse Magnifying Transformer

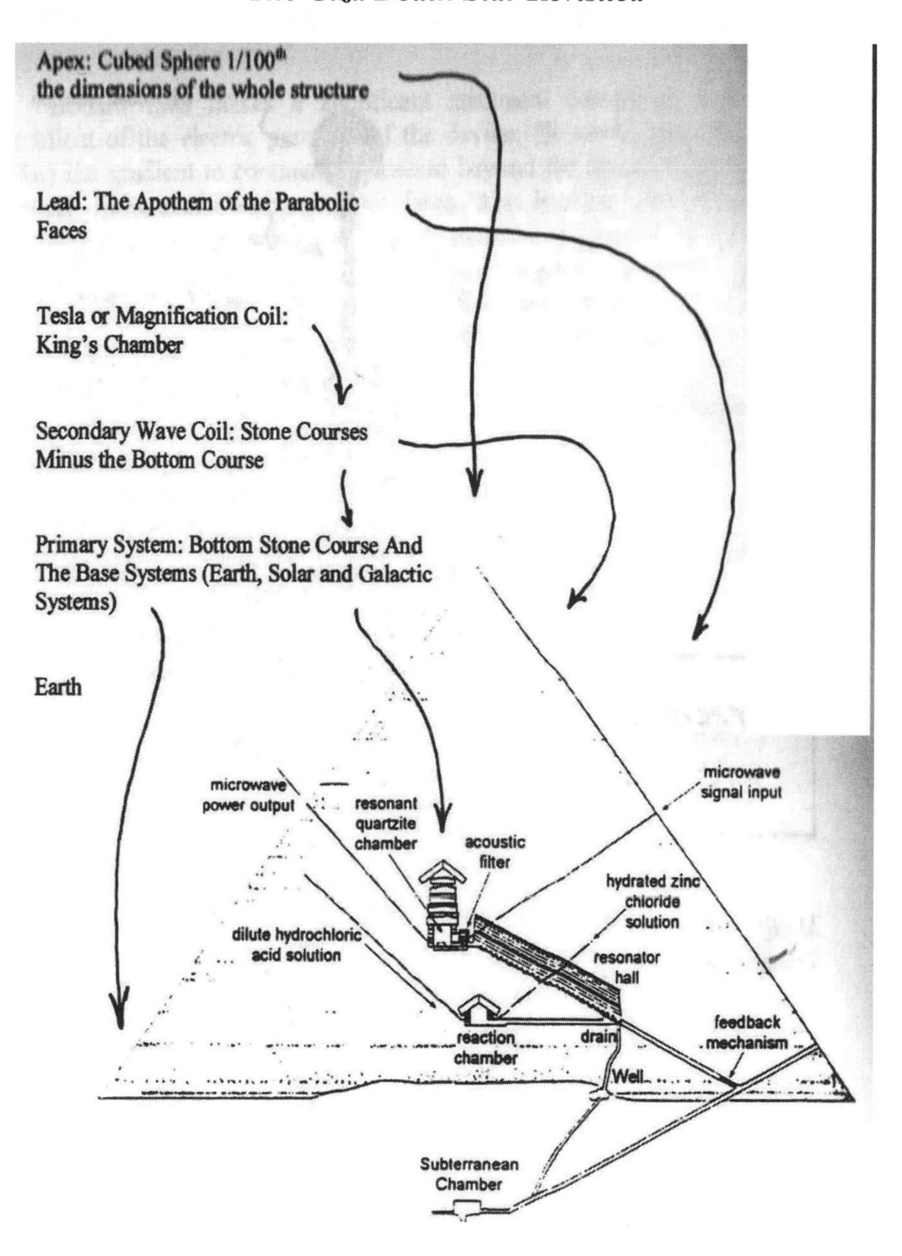

Christopher Dunn's Diagram of the Great Pyramid with a Tesla Impulse Magnifying Transformer Analysis Superimposed

We approach now one of the most significant and difficult points to understand about the Tesla Impulse Magnifying Transformer, and hence, about any attempt to understand how that system might have been functioning with respect to the Great Pyramid.

[Why this is so may not be readily apparent without a consideration of the role of the "ground" in the Tesla magnifying transformer. Dollard's insights here are crucial and fundamental:

> Because the energy is propagated (through) the 'ground' the question exists as to how to ground the apparatus, that is, how to establish an electric reference point, since the so called ground is not the hot terminal of the transponders, and therefore is *incapable of also serving as an electric reference point.* Here exists the singular feature of the Tesla... transformer in that the distributed mutual inductance and odd function resonance work to establish a *virtual ground.... The principle behind this is the geometrical reconfiguration of the fundamental components of energy, the kinetic and potential, this reconfiguration resulting in the separation of cause and effect in not only time but also in space.*[22]

The implications are enormous.

What is normally called the "ground" in electrical circuit schematics... is the Earth, and the antenna **(projects upward from it into)** space, **(and broadcasts into space).** Tesla's impulse technology simply inverted this relationship. *The "ground" became the geometric configuration of space. And the "antenna" became the Earth.*][23]

3. Tesla's Longitudinal Electrical Impulse Waves

Vasssilatos' and Dollard's analyses, and the relevance of the impulse magnifying transformer system to the weapon hypothesis, are confirmed by Tesla's own descriptions of his system. The system associated with Tesla's Impulse Magnifying

---

[22] Dollard, *Theory of Wireless Power*, p. 11.

[23] Joseph P. Farrell, *Giza Death Star Deployed*, p. 203.

Transformer, however, is also that system that Tesla developed and tested in his famous—but little understood—Colorado Springs experiments in the late nineteenth century. For our purposes, it is crucial to remember that up to this point, Tesla had been concerned with alternating current, and with demonstrating the superiority of his polyphase alternating current system to the direct current system of Thomas Edison. Everything changed with the Colorado Springs experiments, for what Tesla had discovered, and would ever after advocate, was yet another entirely new system, one that, in his mind, had many more potential applications than his alternating current system, including, as we shall see, its use as a weapon. Here again, with the reader's indulgence, I want to cite a lengthy section of what I wrote about this system in my book *Babylon's Banksters: The Alchemy of Deep Physics, High Finance, and Ancient Religion.* Again, this citation will begin with a bracket "[" and end with one "]", with new commentary which I have added to the text denoted by **boldface font.**

As I observed in that book, the problem with Tesla's Colorado Springs experiments was that Tesla's analyses and conclusions flew in the face of subsequent "physics orthodoxy". As a result of this, few physicists and engineers examining his experiments really understood what Tesla had discovered nor what he was doing. Those few who *did* understand his claims ended up questioning his [claims, his analysis, or both. The reason for the difficulty in understanding among some scientists, and the questioning attitude and skepticism among others, lies in the nature of what Tesla himself claimed he had been able to do. It is necessary to cite his comments at length:

> Towards the close of 1898 a systematic research, carried on for a number of years *with the object of perfecting a method of transmission of electrical energy through the natural medium,* led me to recognize three important necessities: First, to develop a transmitter of great power; *second, to perfect means for individualizing and isolating the energy transmitted; and third,*

> *to ascertain the laws of propagation of currents through the earth* and the atmosphere.[24]][25]

At that juncture, I asked the reader of the book to [Observe carefully, however, what Tesla states. In order to do what he claimed to do, it was necessary:

1) to construct a kind of transmitter of extraordinary power;
2) to render any wireless transmission of power practical, it was necessary to be able to send and receive a multitude of individual signals for the various equipment and localities it was presumably to serve, and thus, a means had to be found to "individualize and isolate" the transmitted energy, much as a radio receiver can tune to various frequencies to receive different signals; and finally, and most importantly,
3) Tesla clearly states that the "natural medium" itself is to be the means of propagation of this energy.

But what does he mean by "natural medium"? The interpretation of this one point is crucial to the understanding of what he was really seeking in Colorado Springs, and it is this precise point over which the misunderstandings, arguments, and the skeptical questions, arise.

There are three possible ways to understand Tesla here. The first is to understand by "natural medium" what Tesla later declares in the same context: the atmosphere, and the Earth itself. In the first instance, the atmosphere, Tesla is proposing little more than a high-power version of radio. In the second case, however, it is clear that he is proposing something radically different, for the

---

[24] Nikola Tesla, "Transmission of Electrical Energy Without Wires," *Electrical World and Engineer*, March 5, 1904, cited in David Hatcher Childress, ed., *The Fantastic Inventions of Nikola Tesla*, pp. 219, 221, emphasis added.

[25] Joseph P. Farrell, *Babylon's Banksters: The Alchemy of Deep Physics, High Finance, and Ancient Religion* (Port Townsend, Washington: Feral House, 2010, ISBN 978-1-932595-79-6), pp. 131-132.

Earth itself is to be the transmitter **or antenna**. This interpretation, as we shall discover, is that which clearly will emerge from the rest of Tesla's remarks.

However, there is one final interpretation, the most radical of them all, that will loom ever larger as we proceed. Tesla, like many physicists and engineers of his day, was an ardent believe in the aether, that is, in the fact that space-time itself was a kind of ultra-fine matter upon which electromagnetic waves, and hence energy and power, could ride. Tesla, however, appears, unlike most physicists and engineers of his time, to have held the view that this aether had fluid-like properties, i.e., that it had the ability to be compressed or rarefied, that is, to be *stressed* **and thus able to transmit longitudinal waves of compression and rarefaction.**[26] This will emerge as a clear implication of his Colorado Springs experiments.

To see how, we return to his own remarks to understand how he intended to utilize the Earth itself as the transmitting antenna for electrical power:

> In the middle of June, while preparations for other work were going on, I arranged one of my receiving transformers with the view of determining in a novel manner, experimentally, *the electric potential of the globe and studying the periodic and*

[26] One difficulty that many people have with Tesla's views is that they consider the Michelson-Morley experiment, and Einstein's special theory of relativity, to have conclusively disproven the idea of an aether as such. I pointed out in the original *Giza Death Star* trilogy that the French physicist Georges Sagnac, who re-performed the experiment in 1913, did so on a *rotating* interferometer system, arguing that Michelson and Morley had misconstructed their interferometer and had not taken rotating systems adequately into account. Accordingly, his rotating interferometer *did* detect an interference pattern, and hence, the aether. In any case, Einstein himself in a backhanded way admitted to the compressibility of the lattice of space-time in General Relativity by arguing a large mass could bend, twist, or "warp" it. *Tesla was, in effect, arguing that electro-acoustic or electrical longitudinal waves made such space warps much more accessible than the large mass-energy conversions required by relativity.*

> *casual fluctuations.* This formed part of a plan carefully mapped out in advance. A highly sensitive, self-restorative device, controlling a recording instrument, was included in the secondary circuit, while the primary was connected to the ground and an elevated terminal of adjustable capacity. The variations of potential gave rise to electric surgings in the primary; these generated secondary currents, which in turn affected the sensitive device and recorded in proportion to their intensity. *The earth was found to be, literally, alive with electrical vibrations*, and soon I was deeply absorbed in this investigation.[27]

Having discovered that indeed the Earth itself was an electrically dynamic system, Tesla concluded that

> Not only was it practicable to send telegraphic messages to any distance without wires, as I recognized long ago, but also to impress upon the entire globe the faint modulations of the human voice,...

(In other words, Tesla *was* thinking in terms of radio, but with the entire Earth as his transmitting antenna!)

> ...(but) far more still, *to transmit power,* ***in unlimited amounts,*** *to any terrestrial distance and almost without any loss.*[28]][29]

With the phrase "to transmit power, *in unlimited amounts*" the mask is off the nature of the system and its weaponization potential. Indeed, we may understand Tesla's remarks as a kind of aphorism: ***in any electrical system utilizing the planet itself as the resonant cavity for broadcast of longitudinal waves in the medium, the application of the system as a weapon of great destructive potential can never be excluded because it is implied***

[27] Nikola Tesla, "Transmission of Electrical Energy Without Wires," *Electrical World and Engineer*, March 5, 1904, cited in David Hatcher Childress, ed., *The Fantastic Inventions of Nikola tesla*, p. 222. Emphasis added.

[28] Ibid., pp. 226-227, italicized emphasis added in *Babylon's Banksters*, boldface and italicized emphasis added here.

[29] Joseph P. Farrell, *Babylon's Banksters*, pp. 132-133.

by the system itself, without the necessity of any hardware modification. One may equally note something else (also indicated by Tesla in his later writings), *one may also use such a system for propulsion, for a kind of "warp drive".*

Noting that lightning strikes during thunderstorms could produce standing waves in the Earth that would send electro-acoustic shock waves over the surface of the planet, Tesla observed the weakening, and then the gathering strength of such waves as they reflected back and forth over the surface of the planet.[30] A simple illustration will illuminate this point. Imagine thumping a hollow sphere of some metal. The "thump" will be most intense at the actual point of the thump (or lightning strike). The thump will expand in a circular pattern from there over the surface of the sphere, where it will reach its own equatorial point, where the energy of the thump will be at its minimum because dispersed over the widest possible surface area in the system. From there, the wave begins to contract, resulting in more and more energy being present at any given point on the surface, until the thump reaches the antipode of the sphere where the energy is, with but little diminution, exactly that of the original point of the thump.

The Earth, in other words, acted momentarily as a non-linear conductor of infinite resistance, like the busbars in a normal circuit at the moment of circuit closure. At that moment, the electrical current is no longer a Hertzian sine wave *within* a medium, but a longitudinal pulse *upon* it, requiring a wholly *new* kind of electrical circuit, far different from that of ordinary direct current, or for that matter, his own alternating current system. [This new circuit is, he noted,

> Essentially, a circuit of very high self-induction and small resistance which in its arrangement, mode of excitation and action, *may be said to be the diametrical opposite of a transmitting circuit typical of telegraphy by Hertzian or electromagnetic radiations.... The electromagnetic radiations being reduced to an insignificant quantity,* and proper conditions of resonance maintained, the circuit acts like an immense

[30] Joseph P. Farrell, *Babylon's Banksters*, p. 134.

> pendulum, storing indefinitely the energy of the primary exciting *impulses and impressions upon the earth and its conducting atmosphere uniform harmonic oscillations which,* as actual tests have shown, may ***be pushed so far as to surpass those attained in the natural displays of static electricity.***[31]][32]

Yet another clue is afforded here about the enormous energies that Tesla is talking about, for the phrase "natural displays of static electricity" in the context of his remarks means nothing less than the massive thunderstorm that he had observed at Colorado Springs, with its enormous lightning strikes generating standing waves in the Earth.

In any case, the circuit parameters of Tesla's new system were beyond those of his own polyphase alternating current or those of standard radio transmission and reception, [for in this instance, the relationship of ground and transmitter were turned on their heads, the earth, normally the ground in an electrical circuit, became the *transmitter and conductive medium,* and the atmosphere, normally the conductive medium, became the *ground.* Moreover, in this new arrangement, the waves were not Hertzian at all, but *impulses*, that is to say, they were longitudinal waves of stress, of compression and rarefaction, in the Earth itself, **flashing over its and through its surface.**[33]

a. Wardenclyffe and the **Other** *Medium: the Aether*

As I noted in *Babylon's Banksters,* upon the conclusion of his Colorado Springs experiments, Tesla returned to New York City, where he promptly began construction on a prototype of his new electrical impulse system in order to demonstrate its efficiency and flexibility, and one of his backers was, as is now

---

[31] Nikola Tesla, "Transmission of Electrical Energy Without Wires," *Electrical World and Engineer*, <arch 5, 1904, cited in David Hatcher Childress, ed., *The Fantastic Inventions of Nikola Tesla*, p. 227, italicized emphasis added in *Babylon's Bankers,* boldface and italics added here.

[32] Joseph P. Farrell, *Babylon's Banksters*, p. 135.

[33] Ibid., p. 136.

well known, the famous financier and banker J.P. Morgan, whom Tesla had convinced was financing only a kind a telegraphy system. Only when Guigliomo Marconi succeeded in beating Tesla to the punch and transmitting the first trans-Atlantic telegraphic signal—using many of Tesla's own patents to boot—did Tesla reveal to Morgan that his new system was about far more than wireless telegraphy; it was about the wireless transmission of *power*, power which, by Tesla's own published and very public writings by that point of time, was acknowledged to be virtually unlimited.

Tesla, having already bought the land for his prototype, began constructing a large tower, at the top of which was a curious bulb or acorn-like structure with nodes covered in electrodes, and a central metal pole descending far into the bedrock of the Earth below. Upon learning of Tesla's real intentions, Morgan (so the story goes) pulled his financial backing from the project when he learned that he would not be able to "meter" the power and earn money from it. As I also pointed out in that book, however, this is an unlikely explanation since Morgan stood to make *a great deal* of money off of the royalty and licensing agreements of equipment and appliances that would be using Tesla's new system, a potential market that was, quite literally, global in scope.[34] So again, why did Morgan pull the financial plug on Tesla's project?

b. Tesla's Admission of the System's Weaponization Potential and a Speculative Reconsideration of Tesla and J.P. Morgan

I strongly suspect that the real reason for Morgan's sudden turnabout is due to the fact that either he, or someone close both to Tesla and Morgan—Tesla's fellow electrical engineer and theorist Steinmetz for example—had alerted Morgan to those expressions in Tesla's published writings that could be taken as implying a weapons use for the impulse system. Tesla's system was relatively simple and easy to engineer. Anyone with a basic competence in

---

[34] Joseph P. Farrell, *Babylon's Banksters*, p. 137

electrical engineering could do it, and with enough backing the system could be scaled up to enormous and—in Tesla's own words—"unlimited" power. After all, it was a system based on the entire *planet*. It was a system that, by Tesla's own admission and words, could pulse the planet and even conceivably, be used to *split* and destroy it.

As his fortunes declined, Tesla became less and less guarded in his acknowledgement that the new system had a horrific weaponization potential. In 1907, for example, in a *New York Times* op-ed piece, Tesla stated:

> As to projecting wave-energy to any particular region of the globe, I have given a clear description of the means in technical publications. Not only can this be done by means of my devices, but the spot at which the desired effect is to be produced can be calculated very closely, *assuming the accepted terrestrial measurements be correct*. This, of course, is not the case. Up to this day we do not know a diameter of the globe within one thousand feet. My wireless plant will enable me to determine it within fifty feet or less, when it will be possible to rectify many geodetical data and make such calculations as those referred to with greater accuracy.[35]

The relevance of Tesla's observations here to the Great Pyramid weapon hypothesis are quite apparent, for the latter structure, as will be shown later in this chapter, has redundant dimensional analogues of various geodetic and local celestial geometries and measures, and moreover, has a *piezoelectric function* which might have aided in the accurate approximation of such measures. In other words, if viewed as a very sophisticated type of Tesla impulse system, the Pyramid would appear to be capable of targeting an effect anywhere on the planet.

However, a year later, Tesla was much less guarded with respect both to the weaponization of his system, but also with respect to the massive destruction it was capable of causing:

---

[35] Tesla's Wireless Torpedo: Inventor Says He Did Show that it Worked Perfectly," *New York Times*, March 19, 1907, emphasis added.

> When I spoke of future warfare I meant that it should be conducted by direct application of electrical waves without the use of aerial engines or other implements of destruction. This means, as I pointed out, would be ideal, for not only would the energy of war require no effort for the maintenance of its potentiality, but it would be productive in terms of peace. This is not a dream. *Even now, wireless power plants could be constructed by which any region of the globe might be rendered uninhabitable without subjecting the population of other parts to serious danger or inconvenience.*[36]

From this is becomes clear that

> Tesla's wireless transmission of power was *one and the same technology as a horrible weapon of mass destruction.* And this raises an ominous series of questions: Had J.P. Morgan pulled his financial backing of the project not out of greed, but, perhaps motivated by some secret scientific advisor's cautionary warnings, had he pulled his financial backing because he did not wish to see technology with such destructive potential in the private hands of a scientist well known for very odd and eccentric behavior? Were Morgan's motivations ultimately altruistic? Or, conversely, did he want to develop the technology for such purposes secretly… and by means of its potential use, establish a world mastery for himself and his own class?[37]

Whatever the answers to those questions—and I am not inclined to dismiss the possibilities that they present—what *is* evident by now is the strong parallels between Tesla's impulse magnifying system and, as considered thus far, the Pyramid itself:

1) Both systems are *segmented*, and as Tesla noted, the power of the system increases with its segmentation. If the

---

[36] "Mr, Tesla's Invention: How the Electrician's Lamp of Aladdin May Construct New Worlds," *New York Times*, April 21, 1908, emphasis added.

[37] Joseph P. Farrell, *Babylon's Banksters*, p. 138.

Pyramid was a system based upon Tesla principles, then its many segmented components in its thousands of stones can only amplify whatever power it might once have produced;

2) Both systems produce a kind of electro-acoustic wave—the Tesla system by rapid opening and shutting of a circuit, and the Pyramid via piezo-electric effect—producing a longitudinal electrical wave that moves over the surface(s) of the system or circuit, rather than *through* it; I say *surfaces* because if the Pyramid is a sophisticated version of a Tesla impulse magnifying system, then I think it is likely that the waveform moves over *all* the surfaces of the structure, including those surfaces between individual stones. In this respect, the Pyramid is, like the Tesla impulse system, a *wave-guide*, with the wave impulse moving from base to apex. Thus, like our hollow sphere analogy earlier, as the *surface area contracts, the energy of the impulse increases dramatically, especially if the impulse moves over* ***all*** *surfaces within the structure, reaching an almost infinite potential at the point of the apex itself.*[38]

3) Both systems have the planet Earth itself as the fundamental resonant cavity of the system, that is to say, both systems are – minimally – engineering systems of a planetary scale and are thus class I civilizational systems types on the Farrell corollary scale of the Kardashev classification system.[39]

---

[38] For those wishing to delve much more deeply into my thinking here, at the point of the apex one is dealing with a finite aperture approximation, at which point it is possible the longitudinal pulse becomes a vorticular impulse in the medium.

[39] The Kardashev classification system was invented by the Soviet astronomer Kardashev as a means of classifying potential extraterrestrial civilizations by the amount of energy they required to sustain them. A class I civilization requires the energy of an entire planet; a class II that of an entire star; and a class III that of an entire galaxy, to sustain it. My own corollaries to this classification scheme look not at the amount of *energy required* to sustain them, but rather, at the *scale of systems said*

4) Finally, if one allows that the Great Pyramid is a more advance generation of Tesla;s impulse system, both systems, by the nature of the type of physics and engineering involved, are weapons systems. They may be capable of being used for other applications and purposes, but among the purposes and uses that can never be excluded are those that Tesla finally admitted to, and, from the tenor of his admissions, they may have been in his primary intentions to begin with.

However, the systems as thus far reviewed differ in one very important respect:

5) Tesla's system is bound to Earth, whereas it will become evident that the Great Pyramid encompasses not only many terrestrial or geodetic measures and analogues, but also *celestial* ones as well. Tesla has made it clear that the "natural medium" of his system is the planet Earth itself.

If one assumed this was all there was to Tesla's system, however, one would be missing the *most significant* thing that Tesla said about it, one which he, as a physicist and engineer working in the Victorian steam-punk world of pre-relativistic physics, would have had no difficulty in concluding: if there was an aether, then the Earth could resonate with it, and through it could resonate with, and thus affect, other planetary bodies.

---

civilizations are able to manipulate and influence to any degree by means of systems engineering. Thus, in the corollary scheme, a class I civilization is capable of engineering systems of a planetary scale (such as weather, atmospheric properties, and so on); a class II civilization is capable of engineering systems of a *stellar* scale, such as its magnetosphere and so on, and a class III civilization is capable of engineering systems of a *galactic* scale. From the two sets of criteria, it is obvious that a civilization might be capable of engineering systems of planetary and stellar scale, as ours is, and not yet require the energy output of planets or stars, as ours does not.

This is, in fact, what Tesla himself stated, *in that very same article* from April 21, 1908, and his words, while meant to be uplifting and calming, contain an undeniably chilling implication:

> What I said in regard to the greatest achievement of the man of science whose mind is bent upon the mastery of the physical universe, was nothing more than what I stated in one of my unpublished addresses, from which I quote: "According to an adopted theory, every ponderable atom is differentiated from a tenuous fluid, filling all space *merely by spinning motion, as a whirl of water in a calm lake. By being set in movement in this fluid, the ether, becomes gross matter. Its movement arrests, the primary substance reverts to its normal state. It appears, then, possible for man through harnessed energy of the medium and suitable agencies for starting and stopping ether whirls to cause matter to form and disappear.* At his command, almost without effort on his part, old worlds would vanish and new ones would spring into being. *He could alter the size of this planet, control its seasons, adjust its distance from the sun, guide it on its eternal journey along any path he might choose through the depths of the universe.* ***He could make planets collide*** *and produce his suns and stars, his heat and light; he could originate like in all its infinite forms. To cause at will the birth and death of matter would be man's grandest dead, which would give him mastery of physical creation, make him fulfill his ultimate destiny.*[40]

In other words, Tesla had already foreseen that his system could not be restricted simply to the planet Earth and whatever was on its surface. Ultimately, the "natural medium" with which he was playing was the aether or lattice-work of space-time itself.

Once this is seen and admitted, then the Kardashev schemes of classification begin to break down, as we are dealing, literally, with waves of compression in the medium that might both propel us to the planets and stars…

---

[40] "Mr. Tesla's Inventions: How the Electrician's Lamp of Aladdin May Construct New Worlds," *New York Times*, April 21, 1908, all emphases added. Tesla's letter to the *Times* was dated April 19, 1908.

... as well as rip them apart.

B. The Implications of the Dimensional Analogues of Local Space-Time in the Great Pyramid

Thus far we have made mention of the many dimensional analogues of local terrestrial and celestial space; the time has now arrived that we must explore these in more detail, and attempt to divine their functional purpose. The trouble, of course, with all such catalogues is two-fold. Firstly, with respect to the Pyramid, this catalogue is quite extensive, and thus no attempt is here made to include an exhaustive list. My aim is simply to include the "more suggestive" of these analogues, and what they might indicate for the weapons hypothesis and how they are integrated into it. Secondly, the problem with tables of numbers is that they tend to make most people's eyes glaze over and "skip to the summary" section. I've included what I think are enough numbers for people to discern the main point, and yet hopefully not so many that the readers' eyes glaze over.

But what, exactly, *is* the "main point" of these dimensional analogues? What, exactly, is their function? How, so to speak, do they *work*?

Let us go back to chapter one, and revisit the statement of the weapon hypothesis once again:

> *The Great Pyramid is a receiver, transformer, and broadcaster of longitudinal waves—or if one prefer, gravito-electric or electro-acoustic waves – in the medium of local space-time, that is to say, it is a coupled harmonic oscillator of such waves.* In order to receive, transform, cohere and broadcast such longitudinal waves, that is to say, *in order to function as a coupled harmonic oscillator of the medium itself, the structure must reproduce as many dimensional analogues of the local structure of space-time within the structure as it can and function as a phase-conjugate mirror of them. As such, its principal means of "targeting" is reliant upon entanglement and resonance*

(another reason for all the dimensional analogues) and non-locality.

In the original *Giza Death Star* I also noted that "*the Pyramid*, as a coupled harmonic oscillator, *seems constructed of several oscillators nested within the structure in such a fashion as to suggest a set of feedback loops being used to amplify that oscillated energy.*"[41]

If the Great Pyramid is a coupled oscillator to systems in local space-time in addition to being a collector and transmitter of longitudinal waves from and to such systems, then we may deduce the following set of basic functions for the dimensional analogues within the structure:

1) They exist to increase the overall efficiency of the structure;
2) They exist to enable the Pyramid to oscillate the specific system or collection of systems of which they are analogues;
3) Thus they also exist to allow the structure to be tuned to and to target a specific system, or to put it differently, one "aims" the weapon by tuning it;
4) Thus the Pyramid is a collection of oscillators within oscillators, and this function implies the *transformer* function as one analogue becomes a harmonic of another, allowing frequencies—and hence energy—to be stepped up or down, depending on the purpose in view. As will be seen in our detailed analysis below, the nested feedback loops appear to be designed to step up frequencies and hence energies.

With these thoughts in mind, let us look at some general celestial and terrestrial dimensional analogies.

---

[41] Joseph P. Farrell, *The Giza Death Star*, p. ii.

1. A Brief General Catalogue of Some Terrestrial and Celestial Dimensional Analogues within the Great Pyramid

As was seen in chapter five, the Great Pyramid was designed and built using units of measure that appear to closely approximate the British imperial system of measures of length. Indeed, it was a whole series of British surveyors of the Giza compound, and especially the Scottish astronomer Royal Charles Piazzi Smyth, who maintained that this unit of measure, the so-called Pyramid Inch, was equivalent to 1.0011 American inches and 1.0010846752 British inches.[42] In the measures which follow I have followed others in the use of this so-called Pyramid Inch (PI) for measures of length.

One factor almost universally noted by commentators on the Great Pyramid is its embodiment of dimensional and dimensionless universal mathematical constants. For example, the Egyptians, it is thought, had a fractional representation of the constant π as 22/7, a constant embodied repeatedly in the Pyramid. If one divides the perimeter of the Pyramid base by twice its height, one obtains "a result approximating the value of π to five decimal places.[43] This is not the only time that the Pyramid's builders exhibited a mathematical knowledge well in advance of when the standard academic narrative maintains a particular thing—in this case, π —was discovered, for a similar prescience appears to be the case with the fundamental constant, ϕ, or 1.618181.... This constant has a number of strange mathematical properties, exemplified in the following relationships: $1 + \phi = \phi^2$, and $1 + 1/\phi = \phi$. What is unique about this constant is that the Pyramid's builders appear to have worked out that $\pi = 6/5 \times \phi^2$, nor is this all.[44] The east wall of the King's Chamber, for example, has a diagonal of approximately 309 PI, with a length of 412 PI

[42] Rodolfo Benavides, *Dramatic Prophecies in the Great Pyramid* (1974), p. 2. Q.v. Joseph P. Farrell, *The Giza Death Star*, p. 174.

[43] Benavides, op. cit. p. 24, Q.v. Farrell, *The Giza Death Star,* p. 174.

[44] J.P. Lepre, *The Egyptian Pyramids: A Comprehensive and Illustrated Reference* (1990), p. 194.

and a long diagonal of 515 PI, embedding knowledge of the Pythagorean 3-4-5 triangle long before standard historiography maintains they discovered it.

There's still more.

Over the years since Greaves, Piazzi Smyth, and Flinders Petrie published their measurement surveys of the Great Pyramid, many people have simply "played with the numbers" and discovered even more intriguing things about the structure. One of them was Julian Gray, who made one of the more suggestive "cosmological" analogues lurking in the numbers, finding the Gaussian constant of gravitation (k) [expressed trigonometrically as "the reciprocal of the distance between the Coffer and the north of south wall of the King's Chamber, minus one ten-billionth of bottom perimeter of the Coffer…"[45]][46] Fascinating and suggestive as this is, however, it does not take into account Dunn's very persuasive argument that the Coffer is not in its original position. If so, then the implication is that someone moved it, perhaps with some purpose involving the Gaussian constant of gravitation.

There is little doubt, however, that the Pyramid also contains dimensional analogues to local celestial space.

[Robert Bauval and Adrian Gilbert state that "the pyramid was built circa 2450 BC according to star alignment data for the 4 air shafts of the King and Queen's chambers."[47] But the conclusion exceeds the evidence. Given the extraordinary degree of mathematical, physical, and astronomical data evident in the Pyramid's dimensions, it is entirely possibly that it was built much earlier with the knowledge that those shafts would align at that moment. **(In short, celestial alignments of certain structural features cannot constitute an argument for dating. Knowledge of regular celestial mechanics would allow a builder to build a structure long *before*, or long *after*, certain alignments.)** We simply do not know. The point is that alignments at certain periods

---

[45] Julian T. Gray, *The Authorship and Message of the Great Pyramid* (1953, p. 275.

[46] Joseph P. Farrell, *The Giza Death Star*, p. 175.

[47] Robert Bauval and Adrian Gilbert, *The Orion Mystery: Unlocking the Secrets of the Pyramid* (New York: Crown Publishers, Inc., 1994 p. 35.

with certain features of the Pyramid are an insufficient basis to date its construction.][48] In this respect it should also be noted that Robert Bauval also observed that the Giza compound itself, and the placement of the three major pyramids there—Cheops, Cephren, and Menkaure—"are a reflection of the positions of the stars (in the belt) of the constellation Orion circa 10,400 BC."[49]

Yet another set of dimensional measures that has been a favorite of Pyramid researchers—as well as their detractors—is the fact that if one takes the length of a base side of the Pyramid to be a "pyramid cubit" rather than approximately 9131 pyramid inches, the length happens to be about 365.24 cubits, which is, of course, the number of days in a terrestrial year.[50] Each base side of the Pyramid, in other words, is an almost exact temporal measure of the time it takes for the Earth to complete one revolution in its orbit around the sun. Nor is this all, [for it was also built to indicate exactly when the Earth would pass through its solstices and equinoxes,[51] and, at exactly at noon during the spring equinox, due to the precise angles of its faces, it casts no shadow whatsoever.[52]

It is also built in a ratio of the mean distance of the Earth to the sun, for its height times 10^9 equals the mean radius of the Earth's orbit around the sun, the basic astronomical unit.[53] Likewise, if one doubles the perimeter of the bottom of the Coffer, and multiplies it by 10^8, one obtains the mean distance to the moon.[54] One may also find the ratio of the sun's radius expressed as a function of the measure of the perimeter of the Coffer,[55] as well as another peculiar ratio. "The pyramid embodies a scale ratio of 1/43200. The height times 43,200 = 398.685 miles, which is the polar radius of the earth within 11 miles."[56]

---

[48] Joseph P. Farrell, *The Giza Death Star*, p. 176.

[49] Bauval and Gilbert, *The Orion Mystery*, p. 124.

[50] Julian T. Gray, *The Authorship & Message of the Great Pyramid*, p. 5.

[51] Rodolfo Benavides, *Dramatic Prophecies of the Great Pyramid*, p. 9.

[52] Julian T. Gray, op. cit, p. 111.

[53] Rodolfo Benavides, *Dramatic Prophecies of the Great Pyramid*, p. 11.

[54] Julian T. Gray, *The Authorship & Message of the Great Pyramid*, p. 106.

[55] Ibid., p. 267.

[56] Graham Hancock and Robert Bauval, *The Message of the Sphinx* (New York: Three Rivers Press, 1998 ISBN 0-517-88852-1), p. 38.

The most astonishing astronomical feature embodied in it is the precession of the equinoxes. If one measures the distance from the ceiling of the King's Chamber to the apex of the Pyramid, one will get 4110.5 pyramid inches. This is the radius of a circle whose circumference yields a numerical value giving the number of years it takes to complete a precession of the equinox: 4110.5 x 2 x π = 25,827.[57]][58]

There is yet another series of odd relationships to be found in the Pyramid. "There is so much stone mass in the Pyramid that the interior temperature is constant and equals the average temperature of the Earth, 68 degrees Fahrenheit."[59] Amid all the wonderful and weird "numerology" in the structure, this might be the weirdest and most significant, for it means that the mean thermal gradient of the Pyramid is exactly that of the Earth.

[And since the mean thermal gradient of the Earth is the result of a constellation of factors such as distance from the Sun, the amount of radioactivity absorbed from the Sun, its orbital velocity, mean density, electromagnetic field strength, the rotational tilt of its axis **(as exemplified by the Pyramid's knowledge of the precession of the equinoxes)** and so on, the Pyramid thus reflects an extremely accurate **(and sophisticated)** knowledge of solar and terrestrial physics.

The Great Pyramid is also the most accurately aligned building in the world. It is aligned to true north with only 3/60ths a degree of error. Likewise, it is located at the exact center of the surface of the land mass of the Earth, since the east-west parallel and the north-south meridian that both cross the most land intersect at only two places on Earth, one in the ocean and another precisely at the Great Pyramid.[60] That's not all. The average height of land above sea level is approximately 5449 inches, which is also the Pyramid's height.[61] Stop and consider what this means. Not only

---

[57] Rodolfo Benavides, op. cit., p. 22.

[58] Joseph P. Farrell, *The Giza Death Star*, pp, 176-177.

[59] Rodolfo Benavides, op. cit., p. 40.

[60] Rodolfo Benavides, *Dramatic Prophecies of the Great Pyramid*, pp. 71-72.

[61] John Zajac, *The Delicate Balance*, (1989), p. 153.

did the civilization that built the Pyramid have to possess advanced topographical data of the entire surface of the Earth, it would also have had to possess extremely sophisticated mathematical techniques in order to calculate such a measurement accurately. Additionally, in order to embody all these features in one structure, it would **(probably**) have had to possess **(some form of)** computer-aided design and architectural technology of some sort analogous to our own.][62]

2. Stone Courses and Atomic Weights

In addition to this strange catalogue of dimensional or numerological analogies, analogies which one may classify as "celestial" and "geodetic" or "terrestrial", there is yet a third but little known and seldom discussed series of "elemental analogies" mentioned by E. Raymond Capt in his *Study of Pyramidology*, analogies to some of the chemical elements. These analogies are revealed via numerical analogues to their atomic weights in the masonry courses.[63] There are two methods of detecting this correspondence according to Capt:

> ***Method One:***
> It is known that the various elements are grouped according to their characteristics, or their properties. According to one scientific theory, the element Neon (ne) No. 10 is the nucleus of all succeeding elements and the true starting point of the atomic development. By a mathematical process based upon the altitude of Course No. 10 of the (Pyramid's) masonry, we find the atomic weight of Uranium (U) No. 92 corresponds with the 92nd course of masonry above course No. 10.
>
> The atomic weight is arrived at by taking the said altitude in (Pyramid) inches and moving the decimal point one place to the left, thus dividing by 10. (Pyramid course No 92 is 2387 (Pyramid) inches above course No. 10, thus yielding 238.7 as the atomic weight of Uranium No. 92.... By the same method

---

[62] Joseph P. Farrell, *The Giza Death Star*, pp. 178-179.

[63] E. Raymond Capt, M.A., A.I.A., F.S.A. Scot, *Study in Pyramidology* (Thousand Oaks, California: Artisan Sales, 1986), p. 251.

of computing, other elements in the same group show the following relationships.[64]

He then reproduces the following table and diagram:

ELEMENT	ATOMIC NO.	ATOMIC WEIGHT	COURSE NO. PYRAMID	HEIGHT ABOVE COURSE NO. 10	ONE TENTH HEIGHT
URANIUM (U)	92	238.07	92	2387 P."	238.7 P."
THORIUM (Th)	90	232.12	90	2320 P."	232.0 P."
RADIUM (Ra)	88	226.05	88	2259 P."	225.9 P."
RADON (Rn)	86	222.00	86	2214 P."	221.4 P."

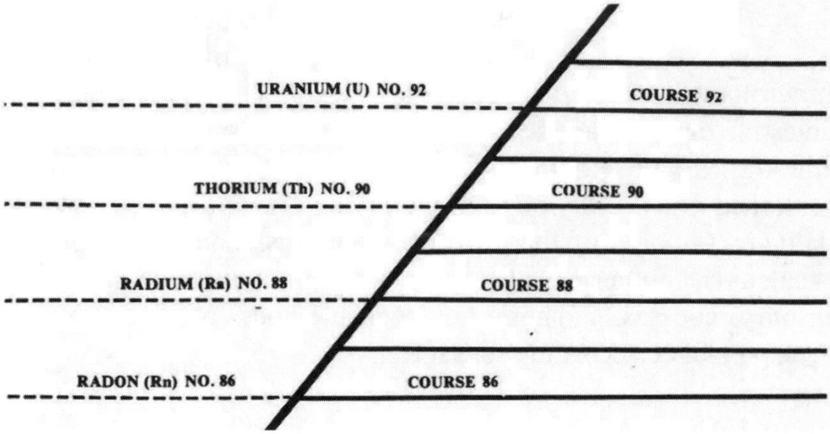

E. Raymond Capt's First Diagram of Atomic Weights and Pyramid Courses[65]

With respect to the *second* method of embodying the "elemental analogues" in the Great Pyramid, Capt states the following:

> Another different but equally significant agreement between the atomic elements and the (Pyramid) course measurements is found by dividing the height (232.512050 Sacred Cubits) of the (Pyramid) in four equal parts. The quarter heights will fall on the following course numbers—43, 95, 152, 215, (Apex). The heights being 58.27 Cubits, 115.80 Cubits, 174.16 Cubits,

[64] Ibid.

[65] E. Raymond Capt. *Study in Pyramidology,* p. 251.

232.52 Cubits. (Apex). The atomic weight of the following elements closely correspond to the four quarter heights:[66]

He then produces the following diagram.

ELEMENT	ATOMIC WEIGHT	NUMBER OF CUBITS ABOVE BASE	
THORIUM (Th)	232.12	232.52	(HEIGHT OF 4TH. QUARTER)
YTTERBIUM (Yb)	173.04	174.16	(HEIGHT OF 3RD. QUARTER)
TIN (Sn)	118.70	115.80	(HEIGHT OF 2ND. QUARTER)
NICKEL (Ni)	58.69	58.37	(HEIGHT OF 1ST. QUARTER)

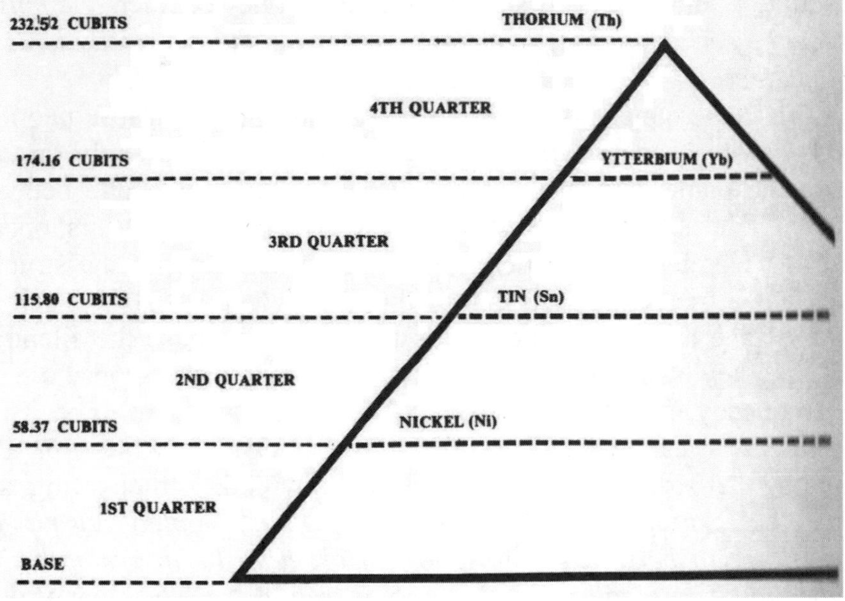

232.52 CUBITS
THORIUM (Th)
4TH QUARTER
174.16 CUBITS
YTTERBIUM (Yb)
3RD QUARTER
115.80 CUBITS
TIN (Sn)
2ND QUARTER
58.37 CUBITS
NICKEL (Ni)
1ST QUARTER
BASE

E. Raymond Capt's Second Diagram of Atomic Weights Embodied in the Pyramid[67]

If true, then Capt's analogues suggest and imply something quite important, something that suggests that the dimensional analogues within the Great Pyramid serve a *functional* purpose beyond merely archiving a system of weights and measures, for

[66] Ibid., p. 252.
[67] E. Raymond Capt, *Study in Pyramidology*, p. 252.

Capt's analogies suggest that one function of the Pyramid is *to covert mass to length*, or to put it differently, that there is a function converting mass to wavelength.

At this juncture, it is necessary to deal with a common objection raised to these sorts of numerical observations. That objection is that the analogues purportedly detected depend upon the accuracy of the set of measurements one is using. If the measurements change, the analogy falls. There is, however, a reply to this argument that can be made from the weapon hypothesis itself: if the Great Pyramid *is* a "coupled harmonic oscillator" of the various systems we have reviewed, then *one must expect a certain latitude in measurements and that this latitude will be deliberate.*

An example may help illustrate this point. Most people know what happens when one presses down a key silently on an acoustic piano, and then strikes the exact same note one octave lower. The silently pressed key will also be heard as its open strings vibrate sympathetically with the struck key. If the struck key has a frequency of x, that of the silently depressed key, being one octave higher, will be exactly x/2.[68] But suppose the silently pressed key is just slightly out of tune, that is to say, suppose it has a frequency of x/2-dx, where "dx" is some increment of x. The *smaller* that increment, the closer x/2-dx is to x/2, and the stronger will be that string's sympathetic vibration with the struck string, x. In other words, *it is not necessary for a coupled harmonic oscillator to be exactly "in tune" in order to be in sympathetic vibration. Close approximation only increases the efficiency of the phenomenon.* And thus, exact measurements are *not* necessary to the oscillator function.

With this said, it is important to observe something *else* about Capt's "atomic weight elemental analogues," and that is that as one goes up the Pyramid toward the Apex, the potential energy implied by the element thus embodied increases. *This is the second example* of a phenomenon mentioned in connection with the

[68] Additionally, the resonator of the sympathetically vibrating string, if composed of the same material as that of the struck string, will be exactly one half the length.

"Tesla impulse magnifying transformer" analysis done earlier: as a standing wave moves through the segmented Pyramid, the segmentation itself amplifies energy, while at the same time, the surface area of such a wave decreases, which *also* amplifies the energy present at a given point.

Later on in this chapter, we will discover yet a *third* indicator that energy is being directed and increased within the overall structure.

3. Oscillators within Oscillators: An Examination of Some of Kingsland's Measures

a. Feedback Loops

But *are* such harmonic functions implied in the Pyramid at all? We have already advanced the thesis that as a coupled oscillator, the Pyramid contains within it oscillators within oscillators, the purpose of which is to construct a series of feedback loops that, in my opinion, act as electro-acoustic transformers, stepping up energy. To see this, one needs only to look at the dimensional measures of certain lengths to be found within the interior chambers of the Pyramid. For this purpose we shall use the measures provided in William Kingsland's *The Great Pyramid in Fact and Theory.*[69]

When one studies Kingsland's tables of measures, it quickly becomes apparent that he has laid out the tables from the outside in, as if one was walking the internal passages and chambers beginning at the entrance of the Pyramid, then continuing through the passageways, into the Grand Gallery, the Antechamber, and finally, the King's Chamber. Laying out the tables in this fashion makes the harmonic relationships between dimensions of these chambers, and hence of the chambers themselves, more apparent. Once one studies these relationships, it quickly becomes apparent that the harmonics and hence the

---

[69] Kingsland's complete table of measures are given on pp. 113-117 of his book, William Kingsland, *The Great Pyramid in Fact and Theory.*

energies of the system are stepped up as one proceeds from the outside of the structure toward the King's Chamber

For example, consider the following table of measurements from Kingsland's list, and it should be noted that Kingsland states the measures in British inches and in an Egyptian cubit of 20.612 inches.[70] Note also that, *for the most part*, I have only analyzed those relationships that are, in a musician's terms, "octaves" or in some relationship to multiples of two, though I've included one or two other relationships to show how deeply the harmonic relationships apply. I have also *numbered* each measure for ease of reference when showing relationships. All measures are in Pyramid inches unless otherwise noted.

Short Horizontal Passage[71]

1) Mean height and width 34.0
2) Mean floor dimensions of the "Recess" 72 x 72

The Ascending Passage[72]

3) Distance of the "point of Intersection" north of the Centre Line 3016.328
4) Distance of the "point of Intersection" south of Base 1518.312
5) Height of the "point of Intersection" above Platform Level 170.696
6) Height of end of Passage above Platform Level 850.396

The Grand Gallery[73]

7) Height of "Step" at the upper end of the Grand Gallery 36.000

[70] Ibid., p. 113.
[71] William Kingsland, *The Great Pyramid in Fact and Theory*, p. 114.
[72] Ibid.
[73] Ibid., p. 115.

8) Height of the Commencement of the Gallery above Platform Level 850.396
9) Height of the top of the "Step" above Platform Level (82 cubits) 1690.184
10) Width of Gallery over the Ramps (4 cubits) 82.448
11) Width of Gallery between the Ramps (2 Cubits) 41.224
12) Width of the Gallery at the roof (2 cubits) 41.224
13) Width of Ramps (1 cubit) 20.612

The Antechamber[74]

14) Height of Passage from Grand Gallery (2 cubits) 41.224
15) Width of Passage from Grand Gallery (2 cubits) 41.224
16) Mean width at floor level (s cubits) 41.224

The King's Chamber[75]

17) Length of Chamber (20 cubits) 412.240
18) Width of Chamber (10 cubits) 206.120

The Queen's Chamber Passage[76]

19) Depth of Step 20.612

The Queen's Chamber[77]

20) Height of Floor above Platform level 834.784

Now let us do a very simple analysis of these numbers, beginning with measures 9-12. Note that these measures are all octaves—multiples of two—and thus are related to each other: 20.612 x 2 = 41.224, 41.224 x 2 = 82.448, which last measure is a

[74] William Kingsland, *The Great Pyramid in Fact and Theory*, p. 115
[75] Ibid., p. 116.
[76] Ibid.
[77] Ibid., p. 117.

close approximation to 82. But what of measurement 9? Kingsland gives the clue when he notes that it is 82 cubits in height above the platform level: 1690.184 ÷ 20.612 = 82. In other words, not only does the Grand Gallery appear to have been designed as an "echo chamber" with specific acoustic properties according to the people who shot pistols inside of it and discovered its ability to amplify and reverberate the sound, but the chamber itself appears constructed of mathematical ratios to enhance that effect. As I noted in the original two *Giza Death Star* books, in effect the chamber is analogous to a gigantic organ pipe.

But does this chamber simply oscillate only to octaves, or are there suggestions it oscillates *other* harmonics than octaves of a fundamental? Answer: yes.

Consider measures 1 and 2 from the Short Horizontal Passage, and compare those to measures 7 and 8 from the Grand Gallery. If one divides the height of the step in the Grand Gallery, measure 7, by the mean height and width of the Short Horizontal Passage, measure 1, one gets 36.0 ÷ 34 = 1.05, almost the same value. Dividing 36 by the Egyptian cubit, 20.612, one obtains 1.7465, or approximately 1.75 or 1 ¾. Following our pipe organ analogy, an organist would call that a mutation.

Measure 2 is the square of 72, or 5,184, which can be divided by the height of the Gallery above platform level, measure 8, to get yet another near whole number multiple: 5,184 ÷ 850.396 = 6.095.

On and on one could go, but it should be apparent by now that the dimensions of the interior chambers are all harmonics of, and hence resonant to, each other, with some dimensions of the King's Chamber acting as very long multiples (by an order of magnitude) of other measures (see measures 16 and 17). Similarly, measures 5 and 6 appear to be closely related, since 850.396 ÷170.696 = 4.9819.

In other words, it is a gigantic coupled oscillator composed of multitudes of coupled oscillators setting up internal feedback loops within the structure. No matter what systems of measure one adopts to view the structure, these relationships appear redundantly, and thus they strongly suggest the structure is meant to vibrate, oscillate, and step frequencies up or down.

b. A Bit of Fun Numerological Speculation About a Very Fine Structure

Even more suggestive results obtain when measures 3-12 are divided by 137, or when 729927 are divided by these measures:

Measure No.	***Measure***	***Divisor***	***Quotient (in some cases, x 19)***
3	3016.328	137	22.01699
4	1518.312	137	11.08256
5	170.696	137	1.24596
6	850.396	137	6.20727
7	36.00	137	2.62773
8	850.396	137	6.20727
9	1690.184	137	6.01810
10	82.448	137	3.00905.
11	41.224	137	3.00905
12	20.612	137	1.50452

But why 137?

137 is the "fractional coefficient" of the fractional approximation of the Fine Structure Constant, 1/137, a *dimensionless* constant or number that—much to physicists' mystification—pops up over and over again in their equations. What is interesting is that using it as a divisor returns nearly whole number results (which in the case of measures 9-12 can be seen when increasing the quotient an order of magnitude, or multiplying by ten).

The same sort of result obtains when one takes the decimal result of the fractional approximation of the Fine Structure

Constant, and then divides it by the same measures. 1 ÷ 137 = .00729927, or a coefficient of 729927:

Numerator	*Measure Number*	*Measure, or Divisor*	*Quotient*
729927	3	3016.328	241.99192
729927	4	1518.312	480.74902
729927	5	170.696	4276.18105
729927	6	850.396	858.33776
729927	7	36.00	20275.75
729927	8	850.396	858.33776
729927	9	1690.184	431.86245
729927	10	82.448	88532.18019
729927	11	41.224	17,706.36037
729927	12	20.612	35,412.72075

This bit of "pyramid quantum numerology" suggests that just as the fraction 22/7 gives a close approximation to the decimal value of π to two decimal places, so such a fractional approximation returns near whole number results when compared to dimensions of the Pyramid in either its fractional or decimal approximations.

This in turn suggests that these appearances in the Pyramid are not accidental and that whoever built the structure may have had a very detailed and thorough analysis of physics and quantum mechanics. If true, then one should expect the coefficients of other constants to be embedded in the structure. In order to examine this "quantum pyramidology" we must turn to the covert and top secret Cold War Soviet research into pyramid power, its astonishing findings, by way of an excursion into American physicist Thomas Townsend Brown, and his experiments and theories in electro-gravitics.

4. Thomas Townsend Brown:
a. Electro-Gravitics: The Impulse System of Brown and its Parallels to Tesla's System

[Thomas Townsend Brown pioneered a little known area of research called electro-gravitics and played a prominent role in the initial planning and design stages of the Philadelphia Experiment to make a ship invisible **(during World War Two).**[78] His patents for electro-gravitic devices are a matter of public record both in England and the United States. They grew from his exploration of a phenomenon familiar to anyone who has worked with cathode ray tubes or toy train transformers. Like Tesla's own transformer, Brown's electrogravitic devices grew from the observation of and further experimentation with sudden high voltage electrical impulses. If a cathode tube is suddenly and energetically switched on, the tube will jerk toward the direction of its positive pole. Electrical energy is suddenly converted to kinetic motion. The phenomenon persists only as long as the impulse, and the tube returns to its rest state just as quickly as it jumped from it. A series of such repeated impulses, Brown reasoned, might conceivably be used as a method of propulsion.

Reasoning that continued impulses of a capacitor could prevent the return to its rest state, Brown constructed a series of experiments as a young man in his garage laboratory. On the basis of his observations, Brown, like Tesla, hypothesized that there was a "form of radiation quite different to the transverse electromagnetic wave. He called it 'radiant energy' and thought that it was present throughout the Universe and was gravitational in nature, but as yet was invisible to instruments."[79] Physicists at

---

[78] For the best introduction to Brown's discovers and how he came to be involved in the alleged Philadelphia Experiment, see Gerry Vassilatos, *Lost Scence* (Kempton, Illinois: Adventures Unlimited Press, 1999, ISBN 0-932813-75-5), pp. 225-281.

[79] Gavin Dingles, "ParaSETI: ET Contact Via subtle Energies," *Nexus,* Vol 8, No 1, January-February 2002, 37-43, p. 40.

Brown's local CalTech rejected his ideas, since it would have meant that gravity was bipolar, repelling as well as attracting.[80]

Initially rejected by university physicists, he came to Kenyon College in Ohio and to the attention of a classmate and friend of Albert Einstein, Dr. Paul Biefeld. Biefeld was captivated by the experiments, and made laboratory space and resources available for more refined versions of the experiments.

> With the new instrumentation and enhanced laboratory access, several details in his strange electric force now became apparent. In 1924, he mounted two spheres of lead on a glass rod and suspended them by two strong insulating supports, forming a swing-like pendulum. When each sphere was oppositely and highly charged with sudden impulses of 120 Kilovolts the entire pendulum swing sideways to a maximum point... and very slowly came back to rest. The electropositive sphere led the motion once again.
>
> What Tom now saw was truly astonishing. The pendulum literally remained suspended in the space for a long time. There were two clearly observable phases in the whole action. The 'excitation phase' took less than five seconds. The 'relaxation phase' require thirty to eighty seconds, coming back to rest *in a series of 'fixed steps'.*[81]

Brown's explanation relied upon Einstein's General Theory of Relativity, which posited space-warping capabilities to large masses exerting enormous gravitational force on the surrounding region of space. The theory further implied a link between electromagnetism and gravity. Brown reasoned that "if gravitation was truly the result of a distorted space then high voltage electric shock was somehow further modifying that distortion."[82]][83]

---

[80] Ibid.

[81] Vassilatos, *Lost Science*, p. 246.

[82] Ibid.

[83] Joseph P. Farrell, *The Giza Death Star Deployed: The Physics and Engineering of the Great Pyramid* (Kempton, Illinois: Adventures Unlimited Press, 2003, ISBN 1-931882-19-3), pp. 205-206.

But why should we bother with Brown's electro-gravitics at all? Thus far, the scientific speculation we have presented considers the Pyramid only from the standpoint of piezo-electrics, though the dimensional analogues of atomic masses has introduced the subject of gravity in a round-about way. There is, however, an important *textual* reason that makes it necessary to include Brown. Recall from chapter three and Sitchin's review of the text of the *Lugal-E,* that Ninurta, when inventorying the "stones" came to the "SHAM" or "destiny" stone, about which Sitchin said this:

> *It was the SHAM ("Destiny") stone. Emitting a red radiance which Ninurta "saw in the darkness," it was the pulsating heart of the pyramid. But it was anathema to Ninurta, for during the battle, when he was aloft, this stone's "strong power" was used "to grab to pull me, with a tracking which kills to seize me."* He ordered it "pulled out... to be taken apart... and to obliteration be destroyed."[84]

The description here is condign to a strong field of gravitation, a "pulling" and "crushing" sensation that military pilots experience in sharp turns or pulling out of a sharp power dive, or that astronauts experience in their centrifuges. In Sitchin's hands, this is being caused by a *crystal*, by a kind of *material.*

This brings us back to Brown, for the essence of Brown's discovery is that of a space-distorting capability accomplished by the exertion of a sudden electrical impulse on a particular kind of material.

Dr. Paul Biefeld, Brown's early mentor, [steered Brown to a series of articles in various scientific journals of the day that recorded experiments where mass was lost in highly charged objects. After all, Michael Faraday himself was aware of the connection: "Electrical capacity is to gravity, as inductance is to magnetism." This led Brown to a further hypothesis: the missing mass was in the distorted *space* surrounding the electro-statically impulse and displaced object.[85] In this, Brown was exploring the

---

[84] Sitchin, op. cit., p. 168, emphasis added.

[85] Gerry Vassilatos*,* *Lost Science*, pp. 246-247.

further implications of the breakdown and inversion of concepts of cause and effect Dollard discovered while experimenting with Tesla's version of electro-impulse technology. In Brown's case, the relativistic order is inverted: mass does not create distorted space, but *distorted space—non-equilibrium conditions—creates the effect mass.* Thrust is therefore provided not by crude thermodynamic processes of action-reaction, but merely by distorting space through high voltage impulse **(*on a non-linear material*, in this case)** a dialectric. Once the space was distorted, the positively charged mass moved into it, and stayed there—apparently defying gravity—until the distortion itself dissipated.

Subsequent experiments with improved capacitors clarified the laws he was uncovering.

> *He found that longer impulse durations required longer relaxation times. Greater dielectric mass in the capacitors amplified the thrusts. Increased voltages amplified the thrust.* He also verified that electric current had nothing to do with the distortion of space at all. Tom estimated the current in these gravitator cells at 3.7 microamps, virtually a 'zero' value. It was the electrostatic impulse which effected the space 'warp.'... Once the gravitator had absorbed the distortion, it stopped accelerating. No amount of additionally applied voltage had any motional effect on the gravitator after this point... Space dynamically interacted in the dielectric with the electrostatic shock.[86]

That is, Brown discovered what Eric Dollard was subsequently to discover **(about Tesla's impulse magnifying transformer):** that the dielectric **(just like Tesla's impulse coil)** functioned as a wave guide, interacting with space itself; only the purpose or use to which Brown put it was different. **(One was interested in the space-warping properties for potential propulsion; the other was interested in the space-warping properties for the transmission of enormous power with little loss or attenuation. *But the underlying system in both cases is remarkably similar*).**

[86] Gerry Vassilatos, *Lost Science*, pp. 247, 248, emphasis added.

Brown further discovered that *the effect was also dependent upon the material composition of the dielectric itself.* Thus, the chemical element or elements, the electrostatic impulse, and spatial geometry exist as a complex of interrelationships in such impulse technologies.

Specifically, Brown discovered that these strange electrical "mass translations" were only detectable if:

1) The K-factor of the dielectric (its ability to store energy) was high (in the order of 2,000 or more);
2) the density of the dielectric was high (in the order of 10 g/cm^3 or more);
3) The applied voltage across the capacitors was high (in the 100,000 V range).[87]

This leads immediately to a consideration of Brown's most profound discovery:

> While working with the gravitator, Dr. Brown *discovered that its behavior as a pendulum varied literally 'with the phases of the moon.'* In addition, there were startling effects which the sun evidently impressed on the gravitator' during its charge-discharge cycles. Whether solar or lunar, it was clear that natural gravitational field conditions were observably affecting local space conditions right before his eyes. The peak maxima and minima of the gravitator varied so much during full moon phases, that he was able to chart the performance against the celestial activities with great precision. After acquiring so much data, he was able to predict what celestial conditions were occurring without visually sighting them. This is when the military became intrigued with his work....
>
> "The gravitator rises during the electrostatic excitation pulse, doing so rather rapidly and discontinuously. When carefully observed, the 'rise' phase consists of several 'graded steps.' Once through this 'stepped rise', the gravitator appears to be in a fluidic channel while suspended at an angle. In this levitated position, that gravitator 'bobs' several times. After the

[87] Gavin Dingley, "ParaSETI: ET Contact Via Subtle Energies," *Nexus*, Vol 8, No 1, January-February 2001 37-43, p. 41.

> shockwave has saturated its dielectric thoroughly, the gravitator begins its lengthy 'fall' beck to the rest point. Here, more than during the rise phase, one most clearly observes the 'rest steps' which last for several minutes.
>
> The discovery identified the number and position of spatially disposed 'rest steps' with the positions of sun and moon. In more refined optical examinations, one could even discern the effect of certain planetary configurations on the gravitator. These fixed space 'slots' became the most intriguing discovery since his original observation of the electrogravitic interaction.[88]][89]

There are several data-points to bear in mind as we proceed, data-points that emerge when compiling together all that we have thus far learned:

1) Electrical impulse phenomena are amplified by circuit parameters, including the number of components or segments in the circuit;
2) Electrical impulse phenomena are analogous to gravitational phenomena in that they are capable of distorting space, and thus manipulating mass itself;
3) Electrical impulse phenomena vary over time, i.e., change as the geometric configuration of local space-time changes. As such, *an efficient system will take these changing circumstances into the design of the system.*

This last point is exactly what we see in the Great Pyramid, and thus it may be safely concluded that the dimensional analogues of local space-time are not *archival*, but *functional* in their purpose.

[Intrigued by this discovery of the solar, lunar, and planetary gravitational effects on his gravitators, Brown was recruited by the Naval Research Laboratory and did highly classified research on gravity wave detection throughout the 1930s. Significantly, this research involved the refrigeration of his units,

---

[88] Gerry Vassilatos, *Lost Science*, p. 251, emphasis added.

[89] Joseph P. Farrell, *The Giza Death Star Deployed*, pp. 208-209.

much like later Soviet **(gravitational)** research involved the use of artificial sapphires cooled to near absolute zero and placed in a vacuum.

Brown's work on dielectrics and gravitation effects soon brought him to a consideration of the materials used in dielectrics. In fact, much of his work for the Navy in the late 1930s involved the investigation of the gravitational and dielectric properties of *granite and basalt*:

> One of the characteristics of a dielectric is its resistivity—how good an insulator it is. If resistivity is not high enough, then the dielectric is rendered inefficient. This is usually a fixed value, but Brown found that the resistivity of some materials would change over time. In a classified naval report entitled "Anomalous Behaviour of massive High-K Dielectrics", Brown described how the resistivity of some materials would alter and even follow sidereal diurnal changes. He also noted that some materials would generate spontaneous radiofrequency bursts whose amplitude was a function of the material's mass and K-factor. Also, he found many granitic and basaltic rocks to be electrically polarized; that is, they behaved like electric cells or batteries. These rocks would have as much as 700mV across them, the amplitude of which would also change in sympathy with solar sidereal cycles. Again, the rock's sensitivity to such changes depended upon its K-factor and mass. It was this latter relation that suggested the phenomenon is gravitational.
>
> During 1937 in Pennsylvania, a Navy-sponsored monitoring station was in operation to record such changes in the electrical self-potential of these rocks. It was noted that there was a strong correlation with the cycles of the moon, which added further support to the hypothesis that the effect is gravitational in nature.[90]

(The relevance of all of this to the Great Pyramid and the weapon hypothesis should be obvious. But) as if this were not enough, there was more, much more, and the attentive reader of the

---

[90] Gavin Dingley, "ParaSETI: ET Contact Via Subtle Energies," *Nexus*, Vol 8, No 1, January-February 2001 37-43, p. 41.

previous chapters will note the significance of the following remarks on Brown's scientific quarrying into the nature of the relationship of gravity and electricity:

> From these investigations, it is clear that the phenomenon is gravitational in nature and that *it manifests as high-frequency electricity.* Brown concluded that the energy is in fact the radiant energy he had hypothesized while still at CalTech.... Constantly emitted from astronomical objects in outer space. While simple high-K dielectric materials would pick up the radiation and convert it directly into electrical energy, *more complex dielectrics such as granitic and basaltic rocks would convert the energy into DC electricity. Not only this, but these rocks are in fact tuned to only a portion of the total radiated energy present throughout the Universe.* This means that your average lump of basalt is a natural gravity-wave AM receiver, tuned into only a few specific radio stations![91]

Let us pause and note what is being said here:

1) High mass increases the efficiency of receiving gravitational energy and converting it to electromagnetic energy in the form of DC current;
2) A high capacitance also increases the effect;
3) High applied voltage across the capacitor also increases the efficiency of the effect;
4) The effect is even detectable naturally in granitic and basaltic rocks;
5) The effect is variable over time as the geometric conditions of local space change; and finally,
6) In their natural state, granite and basalt are only tuned to a narrow band of frequencies of gravity waves.

Obviously, the Great Pyramid fulfills all the conditions that Brown observed, and massively so, save one: the sixth, for the

---

[91] Gavin Dingley, "ParaSETI: ET Contact Via Subtle Energies," *Nexus*, Vol 8, No 1, January-February 2001 37-43, p. 41, emphasis added.

peculiar construction of the Pyramid makes it apparent][92] that the many dimensional analogues of local space serve a functional and not merely archival purpose, perhaps even to the extent of serving the function of modulating a "gravitational signature" into the impulses generated in the structure.

[Brown went further than this, filing a patent in 1953 describing the rudiments of a system for communication via "modulated gravitational radiation."[93] This system **(like Tesla's impulse magnifying system)** involved simple modifications to a standard radio system's antenna. A coil is connected at its base to the output of a high power radio transmitter "so that the radiofrequency energy is end-fed." At the other end of the coil is a spherical electrically conducting body. This acts as "an isotropic capacitor, and so forms a tuned circuit with the coil."[94] The similarities to Tesla's basic system of wireless power transmission will be noted. Perhaps it is equally significant that Tesla, as most researchers into his Colorado Springs experiments are aware, claimed to have received extra-terrestrial communications with his own device.[95]

One should also note that Sitchin maintains that one of the primary functions of the Great Pyramid was as a *communications device* **(in addition to its weapons function)**, and that the structure exhibited strong gravitational pull when fully operational.[96] Could the Great Pyramid, in the light of all of this information, possibly have functioned as either a communications device or power plant? Since the same basic technology and configuration are involved both in Brown's and in Tesla's version of the equipment, the answer would seem to be "yes". At lower power output it could function as either. But it is the Pyramid's over-engineering, its incorporation of analogues][97] of local space,

---

[92] Joseph P. Farrell, *The Giza Death Star Deployed*, pp. 209-211.

[93] Gavin Dingley, "ParaSETI: ET Contact Via Subtle Energies," *Nexus*, Vol 8, No 1, January-February 2001 37-43, p. 41.

[94] Ibid.

[95] Ibid.

[96] Q.v. Joseph P. Farrell, *The Giza Death Star*, p. 50.

[97] Joseph P. Farrell, *The Giza Death Star Deployed*, pp. 211-212.

its coupling to the entire planet as well as its sheer size and mass, the sheer amount of its segmentation, not to mention the textual background, that would [also seem to indicate that it was capable of a much more destructive use, and that such usage was its intended purpose and function.][98]

b. T.T. Brown, The Space Warp, Quantized Macrosystems States, and the Philadelphia Experiment

Townsend Brown was also involved, as previously noted, in the Philadelphia Experiment, and it is in understanding that involvement that more details and date emerge that tie his impulse system and that of Tesla to the Great Pyramid weapon hypothesis. This involvement begins [with a rather unconventional extension of a very conventional technology: arc-welding. The US Navy had constructed a highly classified facility for arc-welding armor-plated hulls.[99] The difference between this facility and normal arc welding was simply its sheer size. The arc for the welds was provided by an enormous capacitor bank, and provided such a huge discharge **(or impulse)** that it was unsafe for the workers to remain inside the chamber when the welding was actually sparked

---

[98] Joseph P. Farrell, *The Giza Death Star Deployed*, p. 212.

[99] While in the original *Giza Death Star* trilogy of books it is not noted, my subsequent investigations into the Nazi Bell disclosed another possibility which I mention here for the first time. As I've noted in my various books covering the Bell, there are aspects of that technology which, if the stories be true, are similar in nature to the impulse technologies of Tesla and Brown, not the least of which are the use of non-linear materials (plasma) to absorb massive electro-static impulses. In fact, there may be a similar connection to arc-welding as well, since it is not a very well-known fact, but nevertheless true, that the two immense German battleships, the Bismarck and her sister-ship, the Tirpitz, had largely arc-welded hulls and armor plate, making them much stiffer and tougher in their ability to absorb punishment than a standard riveted ship with riveted armor. I posit here that the German welding crews may have noticed, and probably *did* notice, similar effects to their American counterparts, and that scientists like T.T. Brown may have been brought in to study the phenomenon.

since the enormous voltage released an intense bombardment of X-rays.

But strange phenomena were almost immediately observed in the new facility, and this prompted an official Navy investigation.

> Phenomena which have no reasonable explanation at all. Researchers examined the site, separately asked workmen to confirm the rumors they were hearing, and watched the process for themselves in the control booth.
>
> What they saw was truly unprecedented. With the electrical blast came an equally intense '*optical blackout*'. The sudden shock of the intense electric weld impulse was indeed producing a mysterious optical blackening of perceptual space, an effect which was thought to be ocular in nature. This intense blackout was believed to be a result of... retinal bleaching, a chemical response of the eye to intense 'instantaneous' light impulse. This was the conventional answer. The more outrageous fact was that the effect permeated the control room, causing 'retinal blackout' even when personnel were shielded by several protective walls.... Careful examination of the effect before the (Naval Research Laboratory) now proved perplexing. First, the 'blackout effect' could be photographed as well as experienced.[100]

This was the least of the Naval Research Laboratory's worries, for workers reported that tools left in the weld room simply disappeared after the impulse. Filming one such weld with deliberately placed objects confirmed the reports; the objects simply vanished.

Conventional explanations were immediately offered that stated the dematerialized objects were being bombarded with intense and sudden X-rays. They were "cooked" apart. But examinations of the weld room after such disappearances produced no trace gases of the missing items. The Navy then turned once again to Dr. Brown, whose gravitational detection work was already well known to them. Indeed, one gains the impression that,

[100] Gerry Vassilatos, *Lost Science*, pp. 254-255.

for the US Navy at least, Dr. Brown was more of an authority on gravity than Dr. Einstein!

In a briefing given to the Navy investigators, Brown surmised that something like a massless miniature black hole was being created by the high voltage sudden electrostatic impulse of the welder.

> Dr. Brown continued to described what was occurring in and around the arc channel. The channel itself was producing its own 'hard' vacuum in stages. Though occurring in atmosphere pressure, the explosive force of the plasma arc had thrust all atmospheric gases out of the arc in its first few microseconds of formation. The full force of the blast was now occurring across a vacuum dielectric. The vacuum actually hindered the complete discharge of the capacitor bank for a few more microseconds, allowing the potential to build beyond those effects observed in weak lightning channels.
>
> It was in a sudden avalanche that the entire discharge occurred across this vacuum space, warping space through an electrogravitic interaction. The interaction was directly related to the voltages, the dielectric column, and the brevity of the impulse.[101] The normal density of inertial space was being instantaneously pierced, the arc literally 'punching a hole' through the continuum.
>
> The explosive vacuum arc set the stage for 'uncommon' observations. Surrounding the intense electrical impulse, space itself was collapsing; space and everything within that space. The strange blackout effect would be expected if all available light was being bent into the arc channel. Incapable of escaping the distortion of space, the blackout effect spread outward. Provided the distortion was intense enough, a specific large volume of space would be 'drawn' in toward the arc channel. The interaction took a few microseconds to effect. There was no escaping its presence.
>
> Furthermore, the blackout would produce various effects in 'successive stages'. At weak levels, one could maintain the blackout effect without noticing any effects on nearby matter.

---

[101] It will be recalled that similar variables were observed by Tesla to have affected his own impulse results.

> There would be an intensity at which significant 'modifications' of matter would be noticed. These would include internal material strains and spontaneous electrical discharges. Provided the blackout effect was 'slow enough,' these material modifications could tear matter apart in an explosion of electrical brilliance.[102]

That is, the particular rise and rest stages of his gravitator *paralleled the quantized intensity of the warp effect* ***(in the arc-welding facility).*** Complete and near-instantaneous production of the effect would simply totally annihilate all matter it touched, leaving not even any "trace gases." A "weaker" and "slower" use of the effect would case the massless and chargeless shock wave to enter the nuclei of atoms and ruthlessly rip them apart in a hugely violent nuclear reaction *that was not a standard chain reaction.* Why this is so is apparent. The shock wave, being massless and chargeless, is a distortion, **(a warp, a longitudinal wave)** in the fabric of local space and time itself, and hence, a distorted geometry literally rips apart... atoms....

Finally, the Navy correctly surmised that at an even weaker "step" or **(impulse),** the effect would be simply to bend light more gently around the warped region rendering anything inside the bubble invisible. This was clearly conceived by the Navy as the first step in gaining the experimental data necessary to control the effect in its more obviously **(weaponizable)** form.][103] *Notably, the system requires no modification of its hardware components to work now in one capacity, and now in another. The only modifications are to the impulse duration, capacitance, and voltage of the pulse.*

To build such a system, especially if it is coupled to the energies potentials of the planetary and local space-time geometries, is ipso facto to build a weapon, even if it can be used at "lower power" for other purposes.

While the many terrestrial and celestial dimensional analogues present in the Great Pyramid argue strongly for their

---

[102] Gerry Vassilatos, *Lost Science*, pp. 258-259.

[103] Joseph P. Farrell, *The Giza Death Star Deployed*, pp. 212-214.

inclusion in the structure for a functional and energetic purpose, the final argument for such a purpose comes from the top secret Cold War era Soviet research into the phenomenon of the mysterious pyramid energies.

5. The Cold War Soviet Era Pyramid Research
a. Particles, Embedding Spaces, Measurements, and "Crystalline" Space
(1) The Physical, Organic, Psychological, Psychometric, and Paranormal Effects

In the Old Testament book of Ezekiel, there is a passage which compares the heavens, and perhaps even space itself, to a crystal: "And the likeness of the firmament upon the heads of the living creature was as the color of the terrible crystal, stretched forth over their heads above."[104]

[For Ukrainian physicist Volodymyr Krasnoholovets and French topologist Michel Bounias, space is a multi-cellular structure; it is, in short, a kind of "crystal."[105] This in itself would not, of course, be too unusual. Other scientists have held similar views of space, and one might be tempted to see in the phenomenon of "galaxy clusters" or "galaxy walls"—vast areas where galaxies seem to be more concentrated—a manifestation of this type of spatial model. It is not here that their unique insights lie. It is rather in the manner in which they *derive* this space,

---

[104] Ezekiel 1: 22

[105] Michel Bounias and Volodymyr Krasnoholovets, "Scanning the structure of ill-known spaces: Part 2. Principles of construction of physical space," *Kybernetes (The International Journal for Systems and Cybernetics)* (2002): Special Issue on New Theories on Space and Time, p. 1. From the abstract: "An abstract lattice of empty set cells is shown to be able to account for a primary substrate in a physical space. Spacetime is represented by ordered sequences of topologically closed Poincaré sections of this primary space... The combination of these morphisms provides spacetime with the features of a nonlinear generalized convolution. Discrete properties of the lattice allow the prediction of scales at which microscopic to cosmic structures should occur."

mathematically, from a physical "nothing." Of course, that too is not unusual in and of itself. "Big Bang" cosmologists have been deriving existence mathematically from a hypothetical cosmic explosion the size of a nano-point for decades, and more recently string theorists have jumped into the picture with a bewildering array of equations to demonstrate that during this initial "event," four dimensions expanded outward in the "Big Bang" while six other dimensions "curled up" inside it. Compared to this Herculian mathematical wizardry, Bounias' and Krasnoholovets' mathematics is simplicity itself, and for that reason, all the more stunning and compelling....

This being said, it is also worth noting that another difficulty attends any presentation of their work such as is undertaken in this context, and that is its relevance to our hypothesis of the Great Pyramid. For this reason, it is best, perhaps, to summarize the Soviet and Russian research in pyramid power itself, and Krasnoholovets' own approach to the subject, in order to provide a convenient backdrop from which to view the theoretical model adopted in Bounias' and Krasnoholvets' papers.

Little at all would be known in the English-speaking world of this Russian research had it not been for the fact that Dr. Krasnoholovets had contacted American bio-physicist Dr. John DeSalvo, who **(ran)** an association of pyramid researchers called The Great Pyramid of Giza Research Association. Most of what is known of this Russian effort is due in no small part to Dr. DeSalvo's efforts to publicize their work in the West. The contact began in January 2001 when Krasnoholovets contacted DeSalvo. Krasnoholovets worked at the Institute of Physics in Kiev, the Ukraine, an institute considered one of the top military research institutions of the former Soviet Union, developing the technologies for Russian "cruise missiles, remote sensing devices, satellites, space station technology, and other military technology." Krasnoholovets informed De Salvo that he and his colleages had carried out research for the last 10 years inside of 17 large pyramids specially constructed out of fiberglass in various locations in Russia and the Ukraine.[106]

As Dr. Krasnoholovets explained, this research had been a broad-based and interdisciplinary effort, conducting experiments in the fields of "medicine, ecology, agriculture, chemistry, and physics." This research was able to document scientifically the changes in "biological and non-biological materials that occur as a result of being placed in these pyramids."[107]

DeSalvo was subsequently contacted by the Russian director of this research, Alexander Golod, **(at that time)** a director in a state defense enterprise in Moscow.[108] Notwithstanding the military connections of Krasnoholovets and Golod, the latter assured DeSalvo that the purpose of building and researching the pyramids was benign.[109]

The pyramids themselves were built at a sharp angle of 73 degrees and were built entirely from fiberglass, using no metal components. That is, the pyramids were constructed entirely from non-linear material. Golod explained that the sharp angle was chosen from "experimental designs that also included the mathematical relationship called the Golden Section."[110] The pyramids, moreover, unlike the structures at Giza, were entirely hollow.[111] The Russian government supported this massive building project, and so committed was it to the idea that Golod was able to persuade the Russian government to allow "a kilo of rocks that had been placed in one of is pyramids on board the MIR space station." Golod's reason for this strange request was that he

---

[106] John DeSalvo, Ph.D., *The Complete Pyramid Sourcebook* (No publication location provided: 1stBooks, 2003 ISBN 1-410708043-0), p. 117.

[107] John DeSalvo, Ph.D., *The Complete Pyramid Sourcebook*, p. 117.

[108] Ibid., p. 118.

[109] Ibid., p. 120.

[110] Ibid., for pictures of these pyramid, see pp. 119-127 of DeSalvo's book.

[111] Ibid., p. 125.

believed the energy fields they produced may help the space station in some fashion.[112]

The institutions reported by DeSalvo to have conducted this research were The Russian National Academy of Medical Sciences; The Ivanovskii Institute of Virology; The Mechnikov Vaccine Research Institute; the Russian Institute of Pediatrics, Obstetrics, and Gynecology; The Institute of Physics in the Ukraine; The Graphite Scientific Research Institute; The Technological Institute of Transcription, Translation, and Replication; The Gubkin Moscow Academy of Oil and Gas; and the Institute of Theoretical and Experimental Biophysics.[113]

Before we get to the actual known results of the Russian pyramid project and its experiments, three general observations should be noted:

(1) The pyramids are deliberately constructed without any linear (metal) components, thus indicating that the Russians already understood something about pyramids as non-linear devices; the sheer size of some of these pyramids would also tend to indicate that they represent the "next phase" of a project initiated after preliminary tests have been done;[114]

(2) While a few experimental results of the Russian research do indicate a possible military application, most of the known results of the Russian research are in areas that are clearly benign; the Russian project, as we shall see, appears to be designed, therefore, to understand as many basic applications of the mysterious "pyramid power" as possible;

(3) Nonetheless, the presence of scientists with clear military connections does tend to suggest that a military application

---

[112] Ibid., p. 118.

[113] John DeSalvo, Ph.D., *The Complete Pyramid Sourcebook*, p. 118.

[114] Golod's remarks to DeSalvo, as recorded in DeSalvo's book, are an indicator that extensive research had probably been previously done as proof of concept experiments before the Russian government decided to invest serious money into the construction of large pyramids. Q.v. *Complete Pyramid Sourcebook*, p. 20.

lurks in the background of this research. Moreover, as noted in point (1) above, it is likely that the preliminary research began long before the actual construction of the pyramids. This is suggested by the fact that little is known about the intellectual history of why this project was undertaken, with all its massive construction expense alone.

With this in mind, we may now examine some of the extraordinary results of the project.

The experiments began in a 36' pyramid, a third of the size of the largest Russian pyramid, an imposing 144' structure.[115] Among the most obvious results of the Russian experiment was the confirmation—in some rather astonishing ways—of effects other researchers had noted in the West: extended life of organic materials and so on.

There were clearly observed physical effects that came off the large 144' pyramid. Russian radar technicians observed "a large column on their radar" yet nothing could be observed visually. The column was "several miles high and about half a mile wide." While uncertain what the column was, the Russians "conjectured it was some kind of ionized column."[116] While this may seem strange, it tends to corroborate the independent research of Dr. G. Pat Flanagan, who... reported Allied warnings to air crews during World War Two operations in the Egyptian desert, cautioning them not to fly over the Great pyramid below certain altitudes, as this would cause their instrumentation to malfunction.[117] The Scientific and Technological Institute of Transcription, Translation and Replication in Kharkov was able to confirm such an ionic column standing above the pyramid to about 2000 meters.[118]][119]

---

[115] John DeSalvo, Ph.D., *The Complete Pyramid Sourcebook*, p.. 124.

[116] Ibid.,p. 139.

[117] G. Pat Flanagan, *Pyramid Power: The Millennium Science* (Earthpulse Press: 1997), pp. 25-26. See also my *Giza Death Star Deployed*, p. 193.

[118] John DeSalvo, Ph.D., *The Complete Pyramid Sourcebook*, 139.

[In yet another experiment, the effect on electric fields was examined. Electric discharges were found to be reduced inside a pyramid and its area was also demonstrably restricted. As DeSalvo observes, it thus "had powerful defensive properties."[120]

But perhaps the most notable effects were on radioactive materials. *Half-lives were altered*, the strength of concrete was altered, *as were the refractive properties of crystals.*[121]][122]

The attenuation of electrical fields inside a pyramid may seem contrary to the weapon hypothesis, but it should be recalled the Soviet era pyramids were entirely empty. We have argued that the piezoelectric effect may have been multiplied many times over due to the segmentation inside the Pyramid, and if the impulse technology analysis be true, the field moves not *within* the structure, but *over its surfaces.* Clearly pyramids give off some sort of energy as the Soviet experiments show. Another possibility indicated by this Soviet discovery of the attenuation of electrical fields inside the structure is that this may have something to do with why the anti-ballistic missile radar in Nekoma, North Dakota looks like a pyramid that is missing its apex. Perhaps this attenuation affected an increase in the accuracy of such radars. The point is, that the resemblance of these modern military structures and the pyramidal shape is hardly accidental.

While other effects of pyramids were recorded by the Soviet scientists, including marked increases in the efficiency of immunoglobin against viruses inside a pyramid, increased crop yields for various seeds that were placed inside pyramids before being planted, and even the regeneration of previously extinct species of wild flowers in the vicinity of their pyramids, the most remarkable thing for our purposes that the Soviet scientists discovered is the variance of *acoustic fields* inside their pyramids, remarkable, because it ties directly to our argument that the

---

[119] Joseph P. Farrell, *The Giza Death Star Destroyed*, pp. 196-201.
[120] John DeSalvo, Ph.D., *The Complete Pyramid Sourcebook*, p. 140.
[121] Ibid.
[122] Joseph P. Farrell, *The Giza Death Star Destroyed*, p. 202.

Pyramid manipulates the power of the physical medium of spacetime *directly.*[123]

[DeSalvo summarizes the Russian results as follows:

> **Distribution of these fields in the pyramid.**
> Center—very strong 9 decibels (near the top of for(sic) the largest Russian pyramid).
> Over the pyramid—very strong and radiates upwards 7-11 decibels (the largest Russian Pyramid).
> Beyond the pyramid—along East-West line the radiation is about 3 times more intense than along the North-South line.
> Below Pyramid- Radiate downward and is over 5 decibels (for large pyramids).[124]

Clearly, alignments to the compass points have something to do with pyramid power, and clearly, the apex of the form is the point of greatest concentration of whatever mysterious field of power is generated by the shape.

Two things must be observed by this list. First, quite obviously *the Russians concluded that whatever field was created, it had quasi-acoustic or longitudinal* ***(wave)*** *characteristics.* Second, in order to measure this field in "decibels", the Russians must clearly have been in the process of developing either a new technology, or new techniques, in order to detect this field intensity....

Clearly, *the Russian experiments demonstrate their prior understanding of the acoustic, non-linear, and hyper-spatial properties of pyramids.*[125]][126]

---

[123] John DeSalvo, op. cit., pp. 130, 136-138.

[124] John DeSalvo, Ph.D., *The Complete Pyramid Sourcebook*, pp. 146-147.

[125] The italicized passages in these two paragraphs were not emphasized in the original *Giza Death Star Deployed.* I emphasize them here because of the importance of the Russian experiments, and Dr. Krasnoholovets' interpretation of them, to the weapon hypothesis argument.

[126] Joseph P. Farrell, *The Giza Death Star Destroyed*, pp. 204-205.

Dr. Krasnoholovets and Michel Bounias explain this mysterious pyramid field, not by appeal to various vorticular physical models as do many other pyramid researchers, but by appeal to an even deeper physics and mathematics, one implying that the field is fundamentally tied to the medium itself, and *to its measurement*:

> [Dr. (Krasnoholovets) explains his discovery of this field by first stating that the Great Pyramid was built to intentionally amplify basic energy fields of the Earth on a subatomic, quantum level. He calls these fields inerton fields or waves and has measured them in model pyramids. He proposes that the Great Pyramid is a resonator of these fields produced by the earth. It would be a new physical field like the electromagnetic or gravitational field. This field is what affects the materials placed in the pyramids...
>
> This inerton field is generated due to friction of moving elementary particles through space. Dr. (Krasnoloholvets) does not believe that space is emptiness like Einstein claims but is filled with a substrate, some kind of an ether, as scientists in the 19th and early 20th (centuries) had believed....
>
> It is hypothesized that atoms of the earth vibrate and interact with the ether generating inerton waves. The Great Pyramid concentrates these waves and is saturated with them.[127]

Thus, Dr. Krasnoholovets has come to similar conclusions as other physicists **(and engineers)** investigating pyramids and the Great Pyramid, namely, that their machine-like properties are due to the fact that *they appear to be resonators and amplifiers of the medium itself*, and this in turn implies an aether or substrate with quasi-hydro-dynaamic-properties, (*hydro-dynamic because longitudinal waves of compression and rarefaction can occur within it, and are directly analogous to the space-and-lattice warping properties of any mass.)* Hence, Dr. Krasnoholovets can speak in terms that suggest a quasi-frictional and quasi-inertial property possessed by this substrate. Clearly, we are far from the static inert aether of 19th and 20th century physical mechanics.

---

[127] John DeSalvo, Ph.D., *The Complete Pyramid Sourcebook*, p. 147.

(2) Topology, the Boolean Lattice of Space and the Quasi-Crystalline Structure of the Medium

Topology is a mathematical language invented in large part by the French mathematical genius Poincaré. For those who do not know what topology is, we shall concentrate on three basic areas in order to understand some aspects of Bounias' and Krasnoholovets' papers and the model that Krasnoholovets uses to explain the mysterious inerton field generated by pyramids: (1) the difference between topology and geometry; (2) basic shapes; and (3) sets and functions.

Topology is basically a kind of geometry without dimensions, or alternatively, a kind of geometry in more than three dimensions. This emphasizes its aspect as *a mathematical technique used to describe objects that exist (or that can only exist) in more than three spatial dimensions*. In fact, one can have as many dimensions as one wishes. The technique was developed since, obviously, it is impossible to draw or model such objects accurately and preserve the characteristics they would have in such higher dimensional spaces. So, think of topology as a kind of "consciousness expanding exercise" with equations. The equations do nothing by draw a *picture* that cannot be drawn in any other way.

The second thing topology concerns itself with is basic shapes, and their similar characteristics. Consider the example that almost every student of topology learns in school: what is the similarity between a doughnut and a coffee cup? For most of us, accustomed to thinking in purely geometric terms, there is little similarity except for the hole in the doughnut, and the hole that is the handle of the coffee cup. But for a topologist, these are the only and primary properties of both.[128] Topologically speaking, then, the coffee cup and the doughnut are one and the same *kind* of object, since the one can be turned into the other by a process of "stretching and pulling" its surface into the shape of the other.

[128] The relevance of this observation for Plato's "mathematicals" and "ideal forms" as well as for Leibniz's *analysis situ* is obvious and will be briefly explored in the appendix of this book.

Only the hole remains. For this reason topology is often called "rubber sheet geometry", since it is a study of *very* basic forms that various objects have; coffee cups can be "stretched and pulled" into a doughnut, and doughnuts can be squished and kneaded into coffee cups. The only rule is that one must retain all basic surface features during the transition, or "mapping" of one to the other, in other words, one is not allowed to plug up the hole of one wile stretching and pulling it into the other.

Now this little rule is what allows the technique of modeling such objects to occur, for certain common features of a surface will appear no matter in how many dimensions an object exists. A familiar example will demonstrate how this works. Take the simplest possible shape one can have in two dimensions, an isosceles triangle. Now we all know from geometry that a triangle has some common properties, no matter what kind of triangle it is. In this case, it's an isosceles triangle, with three equal sides, three "nodal points" and three equal angles. Now, stretch and pull this two dimensional figure into three dimensions—topologists and physicists call this stretching and pulling of an object from one dimensional space into a higher or lower one a "dimensional rotation," but it is nothing more than "stretching and pulling"—and what do you get? A Tetrahedron. But note that the properties of the triangle that was so rotated remain the same: each face of the tetrahedron has exactly the same characteristics as the original isosceles triangle. But some new characteristics have appeared because of the addition of a new dimension perpendicular to the first two. Perform a dimensional rotation of the tetrahedron again from three to four spatial dimensions, and again, the same thing happens: the features of the original triangle are preserved, along with the features of the tetrahedron, and some new ones have been added due to the addition of a fourth dimension perpendicular to the first three.[129] This piling up of perpendicular dimensions and

[129] This technique is indeed one and the same technique that is used to model "imaginary" numbers, which play such a huge role in the mathematical modeling of any electrical circuit. Thus, any electrical circuit is, as the Hungarian electrical engineering genius Gabriel Kron

the resulting morphological changes of basic forms is what some aspects of the topological technique are designed to describe.

For our purposes, it is important to observe that a "system memory" has occurred, since the characteristics of the original triangle which began the process recur over and over again, but in each modified dimensional context the characteristics are modified. The signature of the original triangle remains in the tetrahedron and hyper-tetrahedron, and vice versa, the signature of the hyper-tertahdreon remains in the tetrahedron and the tetrahedron remains in the triangle, by virtue of these common properties that topology expresses mathematically, no matter how many dimensions one is dealing with. We have already encountered this "system memory" in previous chapters, and it will now play a very important role.

So what do Bounias and Krasnoholovets *do* with this marvelous mathematical technique called topology? Basically, they use it to model the structure of space—of the continuum or aether substrate—itself. We must first understand that for them, space is not a "vacuum" or "void" in the conventional sense, though it *is* initially a "nothing", that is, a uniform substrate having no observable distinctions or regions within it. With this in mind, Bounis and Krasnoholovets explain the basic approach of their model in the following way:

> Some necessary and sufficient conditions allowing a previously unknown space to be *explored through scanning operators are* reexamined with respect to *measure theory.* Some *generalized conceptions of distances and dimensionality evaluation are proposed*, together with their conditions of validity and range of application to topological spaces. *The existence of a Boolean lattice with fractal properties originating from nonwellfounded properties of the empty set is demonstrated. This lattice provides a substrate with both discrete and continuous properties,* from

---

pointed out, a hyper-dimensional machine, because it can only be mathematically modeled as such.

> which existence of physical universes can be proved, up to the function of conscious perception.[130]

I have italicized those portions of their initial abstract that are of interest here. Obviously, regardless of one's mathematical background, Bounias and Krasnoholovets have set themselves an immense task. From these two short sentences, one may deduce the entire content of their three-part paper.

Note first of all, that their work concerns itself with the answer to a very basic question: Given that the initial condition of the substrate is one of absolute uniformity, and hence, is also that of physical non-observability or no-thing-ness, what functions would allow one to distinguish discrete regions within such a space? In short, how would differentiation arise? Note that what Bounias and Krasnoholovets have proposed, in other words, is a mathematically formal exposition of **(what esoterist-Egyptologist)** Schwaller De Lubicz **(called the)** "primary scission." Note also that the paper bases these "scanning operators"—which are very simple mathematical functions… – on the notion of *measurement*, that is, on the comparison of any of "the discrete regions of nothing" that are so distinguished.

And finally, observe what the significant result of this procedure is: space, as thus mathematically modeled, turns out to be a "Boolean lattice," that is to say, *it has some properties in common with* ***crystals;*** *it is crystalline-like, or quasi-cellular, in structure.* But it also has some properties in common with logic, or rational thought, that is to say, with *information.* These two ideas—a crystalline structure of information—form the conceptual nucleus of their work.

(And here we approach to an absolutely crucial aspect of their theory and one which, as no other theory does, explains why the Great Pyramid appears to be such an exact replica

---

[130] Michael Bounias and Volodymyr Krasnoholovets, "Scanning the Structure of Ill-Known Spaces: Part I: Founding principles about mathematical constitution of space," (*Kybernetics (The International Journal for Systems and Cybernetics*, 2002, Special Issue on New Theories on Space and time), p. 1, emphasis added.

of—as Newton put it—the universe at large, for) the importance of these conceptions for the idea that the Pyramid was an oscillator of the medium itself is now apparent, for *in order to oscillate the medium, any such oscillator must share common topological characteristics with the local structure of that medium as a crystalline "lattice of information".* Bounias and Krasnoloholovets state this principle more formally: "(Any) property of a given object, from a canonical particle to the universe, must be consistent with the characteristics of the corresponding embedding space."[131] That is, the descent of derivation of any object from the medium preserves the signature of how it was derived from it, **(and any efficient coupled oscillator of that medium will therefore have to be comprised of components preserving a record of that "topological descent". This last point will become crucial when we examine the "missing components" of the Pyramid and their crucial role in the weapon function).]**[132]

This serves to explain why, almost alone of all structures on the planet, the Great Pyramid seems to preserve dimensional analogues implying a system of measurements close to the English imperial system—with its very ancient roots—and why its other dimensional analogues appear to embody a knowledge of quantum mechanics as well, for if "(Any) property of a given object, from a canonical particle to the universe, must be consistent with the characteristics of the corresponding embedding space," then as I put it in *The Giza Death Star Destroyed*, "a profound result obtains for the conception of measurement itself, *since measurement must likewise conform to the derived space"*,[133] or as Krasnoholovets and Bounias put it, "No system of measure can be operational if it

---

[131] Bounias and Krasnoholovets, "Scanning the Structure of Ill-Known Spaces: Part I: Founding principles about mathematical constitution of space," (*Kybernetics (The International Journal for Systems and Cybernetics*, 2002, Special Issue on New Theories on Space and time), p. 2.

[132] Joseph P. Farrell, *The Giza Death Star Destroyed*, pp. 205-210.

[133] Joseph P. Farrell, *The Giza Death Star Destroyed*, p. 210, emphasis added.

does not match with the properties of the measured objects."[134] This is to say "that a standard of measure is in itself and in some respects a coupled oscillator to any measured object"[135] And thus, "an arbitrarily chosen unit of length, for example, having no relationship with the object measured, i.e., not being some dimensional harmonic of it, will accordingly be an inefficient oscillator and an inaccurate measure"[136] topologically.

Behind Bounias' and Krasnoholovets' reliance on the language of topology lurk some crucial insights and implications:

[1) The current mathematical "languages" used to describe the interactions of sub-atomic particles with space is inadequate;

2) It is inadequate *because* it is based on a form of mathematical language where *measurements of distance* ***(and speed)*** or more simply, *vectors* are the primary thing in view;

3) A more adequate way to account for the peculiarities of quantum and even "sub-quantum" mechanics" is via set theory, that is, a mathematical language that compares the properties of systems of sets wherein properties of distance and vectors are only sub-sets of a greater set of properties. Simply put, Krasnoholovets is saying that the fundamental language of physics must change from a *linear* mathematical language—points, lines, planes, vectors and so on—to a *non*-linear language inclusive of such things but not *limited* to them. Hence his emphasis on *information.* Sets of physical properties, on this view, are much fuller description of the "information in the field."

---

[134] Bounias and Krasnoholovets, "Scanning the Structure of Ill-Known Spaces: Part I: Founding principles about mathematical constitution of space," (*Kybernetics (The International Journal for Systems and Cybernetics*, 2002, Special Issue on New Theories on Space and time), p. 2.

[135] Joseph P. Farrell, *The Giza Death Star Destroyed*, p. 210.

[136] Ibid.

Thus... Krasnoholovets introduces the idea that the fundamental relationship between a particle and space itself is harmonic in nature, since a particle, by moving, exhibits inertia and sets up an oscillation in space itself, **(or rather, *is* an oscillation in space itself).** Or as he puts it, "It is the space substrate, which induces the harmonic potential responding to the disturbance of the space by the moving particle" itself that is in primary view.[137]

b. The Soviet Discovery:
The Crucial Area-to-Height Ratio=π/2, and the Great Pyramid

But what has all this to do with pyramids? Krasnoholovets' answer is rather breathtaking, **(and while reading this passage, the reader should also think back to what was said about Tesla's Colorado Springs experiments previously in this chapter):**

> Let A be a point on the Earth's surface *from which* an inerton wave is radiated. If the inerton wave travels around the globe along the West-East line, its front will pass a distance $L_1 = 2\pi r_{earth}$ per circle. The second flow spread along the terrestrial diameter; such inerton waves radiated from A will come back passing distance $L_2 = 4\pi r_{earth}$. The ratio is
>
> $L_1/L_2 = \pi/2$.
>
> *If in point A we locate a material object with linear sizes (along the west-East line and perpendicular to the Earth's*

[137] Dr. Volodymyr Krasnoholovets, "Submicroscopic Deterministic Quantum Mechanic," p. 13. For those of a more technical inclination, this means that the probability wave function ψ of standard quantum mechanics becomes a range that defines a particle's inerton cloud, having dimensions of λ along the vector of movement and 2λ in the transverse direction. Since inerton clouds can interact with each other, the wave structure that results is similar to "ultrasound" which can "destroy, polish, or crush" an object (p. 19).

> *surface) such that it satisfies (the above) relation, we will receive a resonator of the Earth's inerton waves.*[138]

That is, the Great Pyramid, because it is constructed in precisely such a fashion and geometric disposition with respect to the Earth, is a coupled harmonic oscillator of the very inertial properties of the planetary space itself. If there is any doubt that this is what the Ukrainian physicist means, he dispels it immediately:

> Note that the Earth inerton field is also the principal mover that launched rather fantasic quantum chemical physical processes in Egyptian pyramids... power plants of the ancients that has recently been proven by Dunn.[139] This means that the Great Pyramid is fundamentally a coupled harmonic oscillator of gravitational energy itself, since in Krasnoholovets' view such interton waves are "carriers of the inert properties of particles"[140] and therefore are the "real carriers of gravitational interaction."[141]

According to the Ukrainian physicist, pyramids come in three basic shapes, defined by their relationship to the ratio of the side "a" of the pyramid to its height "h". Three shapes emerge:

---

[138] Dr. Volodymyr Krasnoholovets, "Submicroscopic Deterministic Quantum Mechanic," p. 20, emphasis added.

[139] Ibid.

[140] Ibid., p. 21.

[141] Ibid., p. 22.

1) A "sharp" pyramid, where the ratio a/h is less than π/2;
2) The Great Pyramid itself, where the ratio a//h is almost exactly π/2;
3) and an "obtuse" pyramid, where the ratio a/h is greater than π/2.

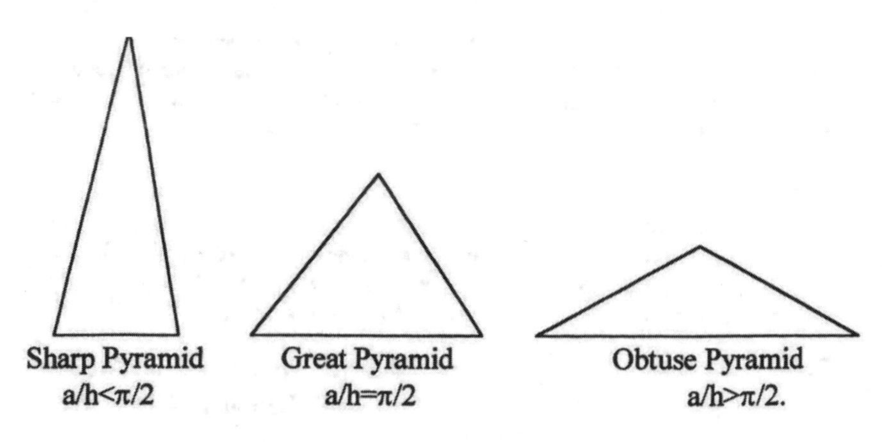

What would the functions of these different pyramidal shapes, as defined by the crucial ration π/2 be? Krasnoholovets speculates that "the sharp pyramid plays the role of a radiator" and that it may also "function as an antenna absorbing inerton radiation from outer space."[142] The obtuse pyramid "to the contrary… may rather function as a radiator that emits amplified inerton waves into the Earth surface."[143] And thus, the most efficient shape to combine both functions would be in the dimensions of the Great Pyramid itself, "the happy medium."[144]][145]

It is crucial to note what Krasnoholovets has actually implied here, for he has implied that the Great Pyramid is designed

---

[142] Volodymyr Krasnoholovets, "On the Way to disclosing the Mysterious Power of the Great Pyramid," p. 14.

[143] Ibid.

[144] Ibid.

[145] Joseph P. Farrell, *The Giza Death Star Deployed: The Physics and Engineering of the Great Pyramid* (Kempton, Illinois: Adventures Unlimited Press, 2003, ISBN 1-931882-19-3), pp. 268-270.

to be both a radiator and collector of terrestrial *and* outer space versions of his "inerton" field; its function, in other words, is not confined to the Earth. Or to argue this differently, its fundamental means of oscillation is via longitudinal waves—what he is calling "inerton" waves and what we have previously called "electro-acoustic" or "electro-gravitic" waves—in the medium itself. Thus, its affects can hardly be confined to the Earth, but rather, can be directed to "any possible target" by means of a kind of non-local harmonic targeting. In such a system of impulses in the medium itself, the Earth is indeed itself the "antenna".

It's not "point and shoot" but rather, "tune and shoot".

In this respect, it is also worth recalling what I wrote about the Soviet pyramid research. Benign though its apparent discoveries appeared to be,

> One has difficulty imagining a former Soviet research institute of physics investigating pyramids solely for medical reasons or the peaceful production of power. Perhaps the Soviet scientists were familiar with the ancient Sumerian texts reproduced by Sitchin that suggested a weapon function for the Pyramid. And perhaps, too, it is significant that Soviet interest in the pyramid power began at approximately the same time that Sitchin published his texts in *The Wars of Gods and Men* in 1987. In any case, the appearance of Krasnoholovets' papers on pyramid power several years after **(U.S. Army)** Lt. Col. Thomas Bearden first raised the warning about Soviet research into scalar weaponry tends to confirm Bearden's analyses, and gives them, to coin a pun, a definite shape.[146]

These considerations—a crystalline lattice of spacetime and longitudinal "electro-acoustic" impulses or warps or waves in that medium—bring us at last to the most difficult aspect of the weapon hypothesis argument, and to what is perforce not only its weakest, but also its most necessary, component: meta-materials and special crystals.

[146] Joseph P. Farrell, *The Giza Death Star Deployed*, pp. 271-272.

C. The Hypothesized, and Missing, Metamaterials to Produce Torsion and Mini-Singularities
1. A Brief Overview of Crystals in Occult Tradition: The Association of Gemstone Crystals and Astrology

Crystals occupy a special place in occult and esoteric lore and tradition. Many people are familiar with one of those traditions, namely, that crystals must somehow be "tuned" to their owner and to the space they occupy. There is a vast tradition within this lore about gemstone crystals and their occult associations, and as these associations form the basis of our speculations on the type of crystal metamaterials that once might have been present in the interior of the Great Pyramid, we must spend a little time reviewing this tradition.

This tradition is widespread, and covers the entire globe; there are ancient Chinese, Hindu, Sumerian, Egyptian, Greek, Roman, Hebrew and even Christian treatises on gemstones, their properties, their influences by (and upon) the planets, people, and the human "humors" or passions.[147] Gemstones could be invested or "possessed" by spirits,[148] and some were said to confer the power of invisibility.[149]

For our purposes, it is the association with planets that is the most important, for it is one of the most widely held beliefs about gemstones that the ancients had. One reason for this is that there was a kind of "color sympathy" at work between the color of the gemstone, the "color" of the human passions or humors, and the predominant color associated with specific planets:[150]

---

[147] See George Frederick Kunz, *The Curious Lore of Precious Stones* (Mineola, New York: Dover Publications, 1971, ISBN 0-486-22227-6), pp. 13-14.

[148] Ibid., p. 5.

[149] Ibid., p. 7.

[150] Ibid., pp. 28-31.

> The magi, the wise men, the seers, the astrologers of the ages gone by found much in the matter of gems that we have nearly come to forgetting. With them each gem possessed certain planetary attractions peculiar to itself, certain affinities with the various virtues, and a zodiacal concordance with the seasons of the year. Moreover, these early sages were firm believers in the influence of gems in one's nativity, - that the evil in the world could be kept from contaminating a child properly protected by wearing the appropriate talismanic, natal, and zodiacal gems.[151]

With the Renaissance, however, came a new attitude, the attitude that will guide us in this last section of this chapter, and which has in fact guided us throughout this entire book:

> … (In) the Renaissance period, an effort was made to find a reason of some sort for the traditional beliefs. Strange as it may seem to us, there was little disposition to doubt that the influence existed; this was taken for granted, and all the mental effort expended was devoted to finding some plausible explanation as to how precious stones became endowed with their strange and mystic virtues, and how these virtues acted in modifying the characters, health, or fortunes of the wearer.
>
> …. Still, when we consider the marvelous secrets that have been revealed to us by science and the yet more wonderful things that will be revealed to use in the future, we are tempted to think that there may be something in the old beliefs, some residuum of fact, susceptible indeed of explanation, but very different from what a crass skepticism supposes it to be.[152]

Later we shall propose a specific and well-known physics reason that may have ultimately been the basis for such traditions, for that same basis underlies our speculation that metamaterial crystals comprised the now-missing components of the Great Pyramid.

---

[151] George Frederick Kunz, *The Curious Lore of Precious Stones*, p. 1.

[152] Ibid., p. 2.

Before we do so, however, we must take note of another aspect of those traditions associating gemstones with astrology and with specific planetary influences. These traditions imply something even deeper, a *cosmological* association of gemstones with the processes, and even the very *beginning*, of the universe itself. For example,

> In Rabbinical legend it is related that four precious stones were given by God to King Solomon; one of these was the emerald. The possession of the four stones is said to have endowed the wise king with power over all creation. As these four stones probably typified the four cardinal points, and were very likely of red, blue, yellow and green color respectively, we might conjecture that the other three stones were the carbuncle, the lapis-lazuli, and the topaz.[153]

For reasons that we shall explore subsequently, it is more likely that the other stones were at least ruby and sapphire for the red and blue color associations, for as we shall discover, carborundum (ruby and sapphire) play a significant role in Jewish crystal lore.

For the present moment, however, it is important to note that this "cosmological" association of gemstones and crystals is not merely an ancient Hebrew preoccupation, for one finds it in Chinese Buddhism:

> The Chinese Buddhist pilgrim Heuen Tsang, who visited India between 629 and 645 A.D., tells of the wonderful "Diamond Throne" which, according to the legend, had once stood near the Tree of Knowledge, beneath whose spreading branches Gautama Buddha is said to have received his supreme revelation of truth. This throne had been constructed in the age called the "Kalpa of the Sages"; *its origin was contemporaneous with that of the earth, and its foundations were at the centre of all things;* it measured one hundred feet in circumference, and was made of a single diamond. When the whole earth was convulsed by storm or earthquake, this resplendent throne remained immovable.

---

[153] George Frederick Kunz, *The Curious Lore of Precious Stones*, p. 78.

> Upon it the thousand Buddhas of the Kalpa had reposed and had fallen into the "ecstasy of the diamond."[154]

Again, this idea that there is a special association of gemstones to the beginning of creation itself is not unique to Buddhism.

One finds it again curiously embodied in an ancient Hebrew tradition; here we must cite the original *Giza Death Star:*

> [According to lore, The Book of the Angel Raziel was inscribed *on a sapphire stone- the stone of destiny*—and was once in the possession of Thoth/Enoch who, it is said, gave it out as his own work,i.e., The Book of Enoch/Thoth. In the legends of the Jews from Primitive Times we learn that, like the Key of Life, *Adam gave the stone to Seth. Who gave it to Enoch, who passed it to Noah, where it develops that Noah learned how to build the Ark by pouring over The Book.*[155]][156]

Here the implication is clear: there is a knowledge or tradition that was passed down from the time of the creation itself.

This implies something quite intriguing for our examination of metamaterial crystals, for *what if the materials, the*

---

[154] George Frederick Kunz, *The Curious Lore of Precious Stones*, p. 238, emphasis added.

[155] William Henry, *One Foot in Atlantis: The Secret Occult History of World War II and Its Impact on New Age Politics* (Anchorage, Alaska: Earthpulse Press, 1998), p. 143, emphasis added. Henry cites Robert Graves and Raphael Patai, *Hebrew Myths* (New York: Anchor Books,1964), p. 113. It is worth mentioning that the psychic Edgar Cayce in some of his visions of Atlantean technology speaks of certain crystals in connection with gravity and destruction: "...(I)n Atlantean land at time of development of electrical forces that dealt with transportation of craft from place to place, photographing at a distance, overcoming gravity itself, prepatation of the crystal, the terrible mighty crystal; mich of this brought destruction." (519-1, February 20, 1934, cited in David Hatcher Childress, *Technology of the Gods: The Incredible Science of the Ancients* (Kempton, Illinois, Adventures Unlimited Press, 2000), p. 296.

[156] Joseph P. Farrell, *The Giza Death Star*, p. 271.

gemstones themselves, as well as the information, from that time were passed down as well?

This "cosmological" association with gems, and especially with blue carborundum (sapphire) is found in Ezekiel 1:26, where the "throne of Jehovah, or the pavement beneath his feet, is compared to a sapphire",[157] and Islamic tradition associates different precious gems with the different heavens.[158] A similar association is found in Hinduism in old texts that indicate how to cast the "nine-gem jewel" that was "designed to combine all the powerful astrological influences."[159] The following are the Hindu correspondences of gems and planets:[160]

Gemstone	***Planet***
Ruby (red carborundum)	Sun
Diamond	Venus
The Moon, Luna	Pearl
Mars	Corak
Saturn	Sapphire (blue carborundum)
Jupiter	Topaz
Mercury	Emerald

The association of sapphire with Saturn is intriguing, for Saturn is, of course, the Greek god Chronos, "Father Time," who waged the "war of the giants" or Gigantomachy with the other gods. As we shall see shortly, sapphire is also associated with gravity and thus according to general Relativity with time. Hence, it would appear that sapphire's association as "the stone of destiny" par excellence

---

[157] George Frederick Kunz, *The Curious Lore of Precious Stones*, p. 275.

[158] Ibid., p. 276.

[159] Ibid., p. 242.

[160] Ibid.

is no accident, and may be a legacy of a forgotten underlying physics.

In raising the possibility that crystals from the beginning of the world might have been handed down as well as information from that time, we return to the cosmological associations that they have with the zodiac. The Jewish commentator Josephus, in his *Antiquities of the Jews*, mentions that the twelve gemstones of the Hebrew high priest's breastplate, or *ephod*, corresponded to the twelve houses of the zodiac,[161] and hence had a cosmological significance associated with the foundation of the world and the beginning of creation. If such crystals *did* exist, they would be objects well worth possessing for reasons that shall be explored in the next section. They would also be objects well worth protecting, and tracking through time. They might also be incapable of destruction, and indeed, there are legends about the indestructible "foundation stone" of the cosmos in much esoteric lore.

In this respect, it is perhaps worth noting what George Frederic Kunz says in his study of gemstone lore about the original breastplate or *ephod* of the Hebrew high priest:

> After the capture of Jerusalem by Titus in 70 A.D., the treasures of the temple were carried off to Rome, and we learn from Josephus that the breastplate was deposited in the Temple of Concord, which had been erected by Vespasian. Here it is believed to have been at the time of the sacking of Rome by the Vandals under Genseric, in 455, although Rev. C.W. King thinks it is not improbable that Alaric, king of the Visigoths, when he sacked Rome in 410 A.D., might have secured this treasure.[162]

If indeed Alaric did make off with the high priest's breastplate, then this would lend credence to those stories that locate the temple treasures in the Languedoc of southern France.

However, there is *another* possibility, one which I have mentioned in many interviews:

---

[161] George Frederick Kunz, *The Curious Lore of Precious Stones*, p. 304.

[162] Ibid., p. 283.

However, the express statement of Procopius that "the vessels of the Jews" were carried through the streets of Constantinople, on the occasion of the Vandalic triumph of Belisarius was placed by Justinian (483-565) in the sacristy of the church of St. Sophia. Some time later, the emperor is said to have heard of the saying of a certain Jew to the effect that, until the treasures of the Temple were restored to Jerusalem, they would bring misfortune upon any place where they might be kept. If this story be true, Justinian may have felt that the fate of Rome was a lesson for him, and that Constantinople must be saved from a like disaster. Moved by such considerations, he is said to have sent the "sacred vessels" to Jerusalem, and they were placed in the Church of the Holy Sepulcre.

This brings us to the last two events which can be even plausibly connected with the mystic twelve gems, - namely, the capture and sack of Jerusalem by the Sassanian Persian king, Khusrau II, in 615, and the overthrow of the Sassanian Empire by the Mohammedan Arabs, and the capture and sack of Ctesiphon, in 637. If we admit that Khusrau took the sacred relics of the Temple with him to Persia, we may be reasonably sure that they were included among the spoils secured by the Arab conquerors, although King, who has ingeniously endeavored to trace out the history of the breastplate jewels after the fall of Jerusalem in 70 A.D., believes that they may be still "buried in some unknown treasure chamber of one of the old Persian capitals."

A fact which has generally been overlooked relative to the fate of the breastplate stones is that a large Jewish contingent, numbering some twenty-six thousand men, formed part of the force with which the Sassanian Persians captured Jerusalem, and they might well lay claim to any Jewish vessels or jewels that may have been secured by the conquerors. In this case, however, it is still probable that these precious objects fell into the hands of the Mohammedans who captured Jerusalem in the same year in which they took Ctesiphon.[163]

[163] George Frederick Kunz, *The Curious Lore of Precious Stones*, p. 283-285.

An equally likely possibility is that if the temple treasures including the breastplate jewels were captured by the Sassanid Persians and their contingent of Jewish soldiers, that the latter would have taken steps to hide and secure those treasure from the advancing Mohammedan Arabs. This may have included *substituting* other stones for the breastplate jewels and allowing the Mohammedans to "capture" them.

In any case, we must now advance a known fact about crystals that might explain this ancient lore of planetary associations, and the traditions of their deeper associations with the cosmos and its beginnings itself.

2. Crystals as Records of History of Local Space-Time Lattice Fluctuations

Let us pause and take stock of what the lore of gemstone crystals says thus far:

1) There is an association of specific crystals to specific planets and their influences, and these crystals are thought to embody such influences in some fashion;
2) There are some traditions of an even deeper sort, associating certain crystals—typically diamonds and sapphires, the two hardest crystals on the Mos scale of hardness—with the foundation or beginning of the world itself; and finally,
3) The implication of the second point is that such crystals from "the beginning of time" or some period in close approximation to it would be objects of power not only worth preserving, but handing down and protecting.

So the important question becomes the following: Is there a scientific fact or theory about crystals that can or could account for these attitudes, and reveal the attitudes themselves as a declined legacy of a forgotten scientific knowledge?

Indeed there is, and in entering this area, we are entering the area of the highest speculation, and highest danger.

In a sense, crystals are "quasi-organic" things, that is to say, *they* ***grow*** over a long time, and grow in a certain way in that *their lattice defects are the direct result of the gravitational circumstances of their formation, and as such, crystals are natural resonators, in their piezo-electric, piezo-acoustic, and gravitational properties, of the topology and geometry in which they grew.* The relevance of this fact to the three points above will be immediately apparent, as will be its power of explanation as to why, in a declining legacy civilization, the effects of this fact were still believed, long after the factual basis for the belief had been forgotten.

Moreover, this is why crystals grown in vacuum space and an extremely low gravity environment have much fewer lattice defects than the same crystal grown in a high gravity environment. Everyday glass, for example, is a crystal that, in the high gravity and atmosphere of Earth has many defects, and hence, it is not used as a load-bearing crystal for construction. In the lower gravity and near-vacuum conditions of the Moon, however, ordinary glass would be much more like steel.

We may extrapolate from this a rather breathtaking series of highly speculative possibilities, namely,

1) that there may once have existed a science of "crystalline descent" that was able to reconstruct a "history" of the geometric disposition of the latticework of the medium itself at a specific time simply by a detailed knowledge of the defect and displacement structure of nodes in the lattice of specific crystals; and hence,
2) this science may also have been able to predict, on the basis of such nodal lattice defects, a probable future disposition of the geometry of the medium; and in either instance,
3) specific crystals will function as efficient oscillators of the medium at specific times, and the closer in time that a

> crystal is formed to the "first time" (or *zep tepi* as the Egyptians called it), the more efficient an oscillator it will be of all *subsequent* geometric configurations, and hence, all subsequent times.

These are the reasons that I wish that I had stated clearly at the time I wrote *The Giza Death Star*, rather than simply assuming that most people would already know them and be able to understand the significance of what I advanced there:

[So what might the Great Pyramid's missing components have been? What was, in fact, once inside the Grand Gallery? I believe that Dunn is essentially correct. They were some sort of acoustic resonators, arrayed in banks that fit into the slots on the side ramps. *But* as Sitchin's **(interpretation of the *Lugal-E* indicated)** they were also much more than these. I believe they were artificial crystals whose crystalline structure *as well as their overall geometric configuration* were carefully—and at great expense—engineered to be both optically and acoustically resonant to the three systems—terrestrial, solar, and galactic—that the Pyramid was coupled to....][164]

In short, the crystals inside the Grand Gallery *were **not**, in my opinion, ordinary crystals, but rather, meta-material crystals with special properties*.

[But this may seem rather fanciful. What could the function of such crystals possibly have been? The solution is simple, but breathtaking], and one that recalls how Sitchin understood the *Lugal-E* as a "crushing" and "seizing" power on Ninurta that "grabbed to kill" him, [*they may have been to resonate to the acoustic harmonics of gravity itself*:

> At Moscow University, Valdimir Braginsky is looking for gravity waves by monitoring tiny changed in the shape of a 200-pound sapphire crystal *cylinder*. *Braginsky chose this exotic material because after being struck it continues to quiver for a record time.* Sapphire's long ringing time permits

[164] Joseph P. Farrell, *The Giza Death Star*, p. 267.

> making a maximum number of measurements before the gravity wave impact fades away. To isolate it from terrestrial noise, the Soviet sapphire is suspended by wires in a vacuum chamber and cooled to near absolute zero....
>
> The first accurate position measurement induces via the uncertainty principle a large momentum spread. For the same reason a collection of particles with different momenta will quickly drift apart, this induced spread in the bar's momentum soon results in a spread in the bar's position. Momentum just happens to be an attribute whose uncertainty feeds back into the position attribute. Braginsky calls such a situation—where accurate measurement of one attribute is spoiled by the back-reaction of the Heisenberg spread in its conjugate attribute—**a quantum demolition measurement.**[165]][166]

But having posited meta-material crystals, what, exactly, might they have been?

[We now enter an area of sheer speculation. Such crystals may have been artificially engineered not so much to refract **(or reflect)** light but to "capture" or absorb light via peculiar lattice properties. Such properties would give these crystals some very peculiar characteristics. I call these artificial crystals "ϕ" or "black crystals." Such "phi" or "black" crystals would palpably resemble black holes and superconductors since such "phi crystals" would also be analogous to a super-conductor, "imprisoning" electromagnetic energy by rotating electromagnetic fields inside of it by ding of its peculiar refractive and lattice properties.][167] Or to put this differently, to *rotate light itself* within the crystal by dint of those peculiar lattice and refractive properties which are beyond those of a negative index of refraction. If one can imagine the facets of a gemstone so arranged to refract and rotate light

---

[165] Nick Herbert, *Quantum Reality: Beyond the New Physics, an Excursion into Metaphysics and the Meaning of Reality* (New York: Anchor Books, 1985), pp. 132-133, italicized emphasis added, boldface emphasis in the original.

[166] Joseph P. Farrell, *The Giza Death Star*, pp. 267-268.

[167] Ibid., pp. 273-274.

completely within the crystal, one approximates to the type of metamaterial being speculated. Such a crystal's surface would, in effect, constitute a boundary condition for a miniature singularity, and the crystal itself, depending on its "color" would appear to be a bluish-black, or reddish-black, or greenish-black, and so on. It would be brilliant, but brilliant in the kind of way of a "field collapse" discussed earlier in connection with Thomas Townsend Brown and the arc-welding facility. Such a rotation of light within a meta-material crystal of this nature [would set up a field in the vicinity of the Pyramid that would literally "pull" or "tug" at anything on the surface or air space around it, exactly what was recorded by Sitchin's texts.][168]

When I originally advanced the notion of metamaterial crystals with such peculiar refractive properties that light itself would be rotated within the crystal like a superconductor rotates electrical current, what I was suggesting seemed remote and far-off, if not far-fetched, particularly with respect to such a phenomenon producing a gravitational pull. But since the time of the original publication of *The Giza Death Star* in 2001, the science of metamaterials has advanced in leaps and bounds, with some of it coming quite close to the conceptions I was advancing. Consider just the following, and it is by no means anywhere close to a complete or even thorough list:

---

[168] Joseph P. Farrell, *The Giza Death Star*, p. 274.

a. Acoustic Metamaterials: Deliberate Lattice Defects and the Reflection and Direction of Seismic Waves

One area of research actively being pursued is that of "acoustic metamaterials,", i.e., materials designed to induce negative indices of refraction. In the case of optical refraction, most of people are familiar with what happens: placing a pencil in a glass of water for example, will make the pencil appear larger, but it will still appear in the water *at the same angle* as it enters the water. But now imagine that the water has a negative refractive index: the pencil will appear *bent in the opposite direction as the angle of entry, and at exactly the same amount of degrees*. An acoustic negative refraction index does something very similar, by reflecting acoustic waves away from the angle of incidence. In other words, "the direction of sound through the medium can be controlled by manipulating the acoustic refractive index... eventually being able to cloak certain objects from acoustic detection."[169]

This feat is accomplished by engineering the *structure* of the material rather than its chemical *composition,* and one of the chief means of doing this is via "the controlled fabrication of small inhomogeneities" in the lattice structure. In other words, acoustic metamaterials are crystalline in structure, and a crucial component of the lattice structure *is the deliberate placement of typical crystalline lattice defects*. This is a crucial point, for in the original *Giza Death Star* series, I actually speculated that the locations of the interior chambers within the Great Pyramid could be viewed as deliberately placed mega-defects in the overall lattice work of the stones comprising the structure.

Obviously, avoidance of acoustic detection would be of great military value in aiding submarines to avoid sonar detection and ranging. But there is another and even more sinister implication, for "Applications of acoustic metamaterial research

---

[169] "Acoustic metamaterial," *Wikipedia*, https://en.wikipedia.org/wiki/Acoustic_metamaterial, p. 2.

include seismic wave reflection and vibration control technologies related to earthquakes."[170] *Reflection* of seismic waves and "vibration control" technologies implies not only the ability to *damp and disperse* such waves, waves which, let it be noted, are longitudinal waves of compression and rarefaction by their very nature, but also implies the ability to *amplify, concentrate, and direct* such waves.

In short, such acoustic metamaterials imply the weaponization of earthquakes. If such were present in the resonator arrays within the Grand Galllery, then one possible use is precisely the damping, or amplification, and directing of earthquakes.

b. Gravity Crystals and Gravity Wave Detectors

Yet another application of such acoustic metamaterials comes in the form of *gravity wave detectors*. In a remarkable article at Phys.Org titled "A tiny crystal device could boost gravitational wave detectors to reveal the birth cries of black holes" by David Blair, a problem with current gravity wave detectors is summarized: they simply do not operate at high enough frequencies to detect many such waves. But there is a possible solution:

> Two different kinds of quantum packets of energy are involved, both predicted by Albert Einstein. In 1905 he predicted that light comes in packets of energy that we call *photons*; two years later he predicted that heat and sound come in packets of energy called *phonons*.
>
> Photons are used widely in modern technology, but phonons are much trickier to harness. Individual phonons are usually swamped by vast numbers of random phonons that are the heat of their surroundings. In gravitational wave detectors, phonons bounce around inside the detector's mirrors, degrading their sensitivity.

[170] "Acoustic metamaterial," *Wikipedia*, https://en.wikipedia.org/wiki/Acoustic_metamaterial, p.4.

Five years ago physicists realized you could solve the problem of insufficient sensitivity at high frequency with devices that *combine* phonons with photons. They showed that devices in which energy is carried in quantum packets that share the properties of both(sic) phonons and photons can have quite remarkable properties.

These devices would involved a radical change to a familiar concept called "resonant amplification". Resonant amplification is what you do when you push a playground swing: if you push at the right time, all your small pushes create big swinging.

The new device, called a "white light cavity", would amplify all frequencies equally. This is like a swing that you could push any old time and still end up with big results.

However, nobody has yet worked out how to make one of these devices, because the phonons inside it would be overwhelmed by random vibrations caused by heat.

In our paper, published in *Communications Physics*, we show how two different projects currently under way could do the job.

The Niels Bohr Institute in Copenhagen has been developing devices called phononic crystals, in which thermal vibrations are controlled by a crystal-like structure cut into a thin membrance. The Australian Centre of Excellence for Engineered Quantum Systems has also demonstrated an alternative system ***in which phonons are trapped inside an ultrapure quartz lens.***

We show both of these systems satisfy the requirements for creating the "negative dispersion"—which spreads light frequencies in a reverse rainbow pattern—needed for white light cavities.

Both systems, when added to the back end of existing gravitational wave detectors, would improve the sensitivity at frequencies of a few kilohertz by the 40 times or more needed for listening to the birth of a black hole.[171]

---

[171] David Blair, "A tiny crystal device could boost gravitational wave detectors to reveal the birth cries of black holes," *Phys.Org*,

Could the Pyramid have functioned as a giant lens of, i.e., as an amplifier for, gravity waves?

If such metamaterials formed the composition of the resonator arrays inside the Grand Gallery, then the answer is "yes," because as we shall see, there is a profound clue present in the structure itself, a clue *other* than its evident "segmentation," which as we discovered in our review of Tesla's impulse system, *amplifies* the force and effect of the impulse. There is present in the structure something *else* that indicates a lens-like magnification function. And in any case, tiny quartz crystals—billions, trillions, quadrillions, quintillions of them—are embedded in the granite and limestone of the structure.

There is yet another association of crystalline structures to gravity, and again, it links directly to the Great Pyramid, particularly if the proposition of Chris Dunn is true and that the Pyramid produced a hydrogen plasma in the chambers of its interior. In yet another article title "Gravity crystals: A new method for exploring the physics of white dwarf stars," a simple *structural* modification of ordinary everyday items is able to create a "gravity crystal":

---

https://phys.org/news/2021-02-tiny-crystal-device-boost-gravitational.html, pp. 2-4, bold-italics emphasis added.

c. Stopping Light Inside of Crystals

At the time I wrote the *Giza Death Star* and postulated the existence of metamaterial crystals that were able to "confine" light within them, I did not imagine that within a few years, engineers would actually be able to do it:

> In a vacuum like space, the speed of light is just over 186,280 miles per second. Scientists have now shown it's possible to slow it down to zero miles per second without sacrificing its brightness, regardless of its frequency or bandwidth.
>
> A team of researchers from the Israel Institute of Technology and the Institute of Pure and Applied Mathematics in Brazil discovered a method of theoretically bringing the speed of light to a halt by capitalizing on "exceptional points"—coordinates at which two separate light emissions reach each other and merge into a single one, according to *Phys.Org.* A paper describing the research was published in the scientific journal *Physical Review Letters.*
>
> Slow-light technologies could help improve our telecommunications systems, as well as our quantum computers. Existing research shows us that light can be slowed to an infinitesimal fraction of its vacuum speed in two ways, according to the paper: trapping it inside either ultracold atom clouds *or inside waveguides made with photonic crystals.*
>
> .... (It's) the second method that allowed the researchers to make their breakthrough.
>
> Photonic crystals are materials with billions of tiny holes through which light refracts ...A waveguide, meanwhile, is a confining tube-like structure, which, as the name suggests, guides the waves sent inside it (any kind of waves, but in this case optical waves).
>
> The problem is that the process of slowing down light tends to sap its intensity. What the team discovered is that such losses can be eliminated if a waveguide is designed with

> parity-time symmetry, a relatively new concept that refers to maintaining a constant balance, or symmetry, between a system's energy losses and gains....
>
> By tweaking the gain-loss parameters, the technique could be adapted to light of any and all frequencies and bandwidths....[172]

Another article from the Massachusetts Institute of Technology clarifies this procedure, revealing a frightening implication:

> Light can usually be confined only with mirrors, or with specialized materials such as photonic crystals. Both of these approaches block light beams; last year's finding demonstrated a new method in which the waves cancel out their own radiation fields. The new work shows that this light-trapping process, *which involved twisting the polarization direction of the light- is based on a kind of vortex*—the same phenomenon behind everything from tornadoes to water swiling down a drain....
>
> But in the case of these light-trapping crystals, light that enters the material becomes polarized in a way that forms a vortex... with the direction of polarization changing depending on the beam's direction.
>
> *Because the polarization is different at every point in this vortex,* ***it produces a singularity—also called a topological defect...at its center, trapping the light at that point.***[173]

With the mention of "singularity" the mask has come off the otherwise bland and benign language of "tired" or "slow" light, quantum computing, and telecommunications, for the term

[172] Kastala Medrano, "Physics: Speed of Light Could Be Brought to a Complete Stop by Trapping Particles Inside Crystals," January 2, 2018, emphasis added.

[173] David L. Chandler, "Trapping Light with a Twister," *MIT News*, December 22, 2014, https://news.mit.edu/2014/trapping-light-minature-particle accelerators-improved-data-transmission-1222, all emphases added.

"singularity" is another name for the extreme gravitational anomaly known as a black hole.

In other words, trapping light within a crystalline structure where the index of refraction and/or polarization is constantly changing in a vorticular structure will produce a mini-singularity or miniature black hole, complete with its own strong local gravitational field. One may imagine such a crystalline structure by imagining the facets of a gemstone, polished and carved to refract and reflect light, and mechanically spinning the gemstone so fast that as the light is refracted and reflected within the stone between the faces of its facets, the light for the most part remains confined within the stone, with little of it "leaking out."

Speculating further, it is conceivable that an array of such crystal resonators in the Grand Gallery would be resonators and oscillators of the harmonics of some gravitational wave fundamental.

But is such a gravitational link of these micro-sized singularities possible?

d. Time-Crystal Ground States

As I have written elsewhere, when CERN's Large Hadron Collider was going to be turned on for the first time, there was some concern among some scientists that it might create a mini-singularity called a quarck-gluon plasma, and that if this occurred, this mini-black hole would be impossible to shut down, and that it would eventually destroy the Earth by literally consuming or "eating" it.

In a rather remarkable paper published online on April 19, 2019, and sponsored by the University of Lund, Sweden, and the Fudan University of Shanghai, China, scientists Andrea Addazi, Antonino Marciano, and Roman Pasechnik proposed that such mini-singularities or quarck-gluon plasmas were natural resonators of the original big bang and the gravitational waves therefrom. With a paper deliberately incorporating "primordial gravitational waves, homogeneous gluon condensate, effective Tang-Mills theory, QCD (quantum chromo-dynamics) phase transition" as its keywords,[174] the paper is a typical physics paper, consisting of a blizzard of differential and tensor equations with more appendices than there is actual paper.

There is, however, one passage that is relatively equation-free, and whose implications are not only enormous, but "cosmologically nihilistic", if one may so put it:

> Since the gravitational waves emission is related to the time variation of the energy=density, it turns out that the (gravitational wave) spectrum is actually suppressed. Most of the trace tensor variation is provided by the pressure component p_u. *The pressure kinks can be efficiently transmitted to the primordial plasma, since the gluonic consdensate its strongly interacting with it.*[175]

Notice what is actually being said here: all singularities are resonant to each other because they have similar characteristics: a gluon condensate is analogous to the "primordial plasma" because effectively both are the same type of structures—vortices—that are doing the same thing—producing strong gravitational waves.

With this, let us pause and take stock of what the implications of these developments are;

1) In the original *Giza Death Star*, I postulated the existence of metamaterial crystals that I called ϕ, phi, or black crystals, whose indices of refraction rotated light within them, producing an optically and acoustically resonant object with gravity itself;
2) Acoustic metamaterials are capable of indices of refraction that can direct acoustic waves in such a fashion as to make an object acoustically invisible;

[174] Andrea Addazi, Antonino Marciano, Roman Pasechnik, "Time-crystal ground state and production of gravitational waves from QCD phase transition," *Chinese Physics C,* Vol. 43, No 6 (2019) 065101.

[175] Ibid., emphasis added.

3) Similarly, metamaterial crystals can also trap light in a kind of vorticular "facet" producing a miniature singularity, analogous to a quarck-gluon plasma, which in turn is analogous to a black hole. That is to say, trapped light in such a crystal might conceivably produce a gravitational effect that would be capable of amplification to such a degree as to be able to cause a warp or compression-rarefaction region of space time itself.
4) Longitudinal waves in the medium are exactly analogous to such gravity waves, which are areas of compression and rarefaction in the crystalline lattice structure of space itself. Such longitudinal waves are produced by high voltage impulse breaking upon a barrier which its momentarily infinite in resistance. As such, an electrical wave is formed that moves super-luminally *over the surface* of a coil rather than *through* it. In such a system, the coil functions more as a wave-guide than a coil as such, and the amount of segmentation of such a system is directly proportional to the amount of amplification of the impulse; the more segments, the stronger the impulse.

From these for simple considerations, it should be readily apparent that the Great Pyramid could function as such a system provided its "missing components" were not only Hemholtz resonator arrays as Dunn speculated, but optical-acoustic metamaterial crystals resonant to and oscillators of gravity itself. It should be abundantly clear by this point that one cannot build such a system without being aware of its potential use as a tremendously powerful weapon, regardless of any other purposes for which it might also have been intended.

3. An Intriguing Tangent: Crystals and the Alchemy of High Pressure: McCarthy, Monmouth, Synthetic Diamonds, Diamond Anvil Presses

Our speculation concerning metamaterial crystals within the Grand Gallery is not yet complete, for one of the strange things about the "stones of destiny" that Sitchin believed once occupied the Gallery was not only that they were removed, inventoried, and destroyed, but that some of these stones were hidden *because they could not **be** destroyed.*[176] As has been seen in *this* book, the possibility arises that some of these stones were formed close to or shortly after the moment of creation itself, in the extraordinarily high pressure and heat gradients of the Big Bang. These stones are already a kind of metamaterial crystal, a kind of "metadiamond" whose extraordinary light-trapping and singularity-mimicking properties might have been the result of the high pressure environment in which they were formed.

The high pressure environment in which such materials are formed are analogous to the high pressures in which the hardest substance known to man—diamond—is formed, even though diamonds are formed at much *lower* pressures than those involved in the Big Bang and its "primordial plasma." Such high pressures involve an alchemy of their own, an alchemy that is perhaps able to create metamaterials with strange properties.

Such was the case when scientists realized that diamonds could be used in special presses to create extraordinarily high pressures to make and create new types of materials, beginning with synthetic diamonds themselves. This led to the creation of the diamond-anvil press in the late 1950s and early 1960s, which consisted literally of two diamonds whose points were brought

---

[176] Q.v. Joseph P. Farrell, *The Cosmic War: Interplanetary Warfare, Modern Physics, and Ancient Texts* (Kempton, Illinois: Adventures Unlimited Press, 2007, ISBN 978-1-931882-75-0), pp. 204-233, Cahpter 8: "The Story of the Stones: A Little More Why."

together against each other, and subjected materials sandwiched between these diamonds to pressures of several hundred thousands of atmospheres.[177]

One interesting feature of this research is that it was being conducted in part at the U.S. Army's laboratories at Fort Monmouth, New Jersey during the 1950s, and at a time when U.S. Senator Joseph McCarthy (R-Wisconsin) was conducting his committee hearings on security breaches at the facility. It is interesting to note that by 1959, two Fort Monmouth scientists, Armando Giardini and Lieutenant John Tydings, were involved in creating the first synthetic diamonds at the facility. As of this writing, I am unaware if Lieutenant Tydings was any relation to Maryland US Senator Millard Tydings (D) with whom McCarthy clashed in 1950 in the infamous Tydings Committee hearings.[178]

4. The Pyramid as Crystal, Coil, and Wave Guide, and the Stone Courses as Virtual Windings

Once the connection of the impulse system to gravity is grasped, then one can discern the analogues between the Pyramid as such a system, and Tesla's impulse system.

[Viewed in one sense, the granite **(and limestone)** core of the Pyramid, its vast stone courses and the geometric shape of the Pyramid itself as a "squared circle" **(and cubed sphere)** would function as a **(giant)** capacitor reliant upon the piezo-electric properties of the granite **(and limestone themselves).** However, viewed in yet another sense, The Pyramid, precisely as a squared circle **(and cubed sphere)** is *an electrical coil that is segmented—exactly in accordance with the principles discovered by Tesla—not only into separate "windings" in the stone courses, but each of these "windings" in turn is segmented into a discrete number of*

[177] See the excellent review of high pressure materials engineering and diamond-anvil presses by Robert M. Hazen, *The New Alchemists: Breaking Through the Barriers of High Pressure* (New York: Random House/Times Books, 1993, ISBN 0-8129-2275-1), p. 204. See also pp. 180, 182, 198.

[178] Ibid., p. 142-143.

stones.[179] The pyramidal form itself gives the distinctive geometry and properties **(not only**) of a Tesla impulse coil **(but of a wave guide)]:**

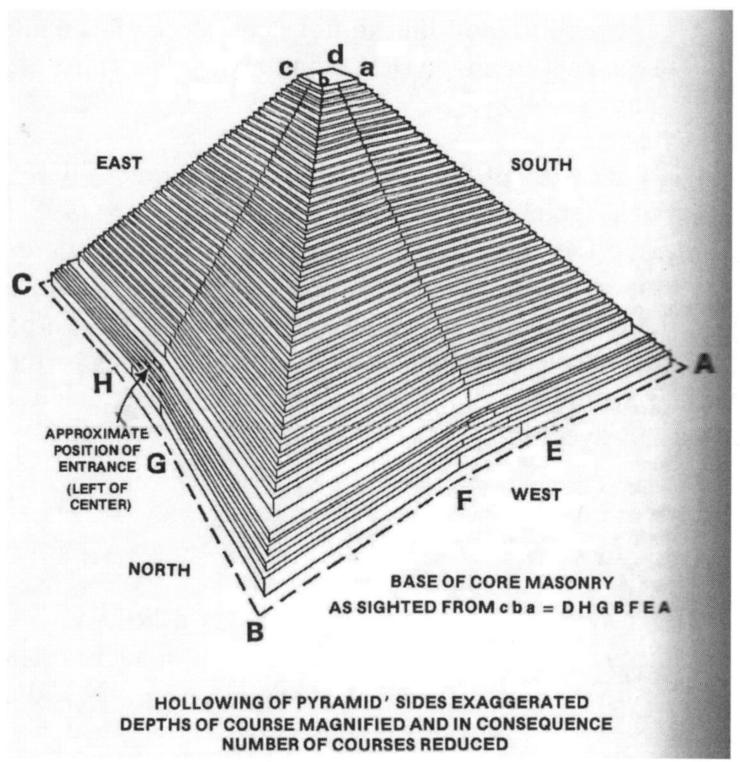

E. Raymond Capt's Diagram of the Interior Stone Courses, Inadvertently showing its overall waveguide shape.[180]

---

[179] And like everything else in the Pyramid, the exact number of stones in each course or winding is probably the result of an exact mathematical and physical design, **(though this aspect of its design has never been suggested nor estimated by any study. I posited this point in the original *Giza Death Star* based on the presence of the atomic weight analogues mentioned by Capt.)**

[180] E. Raymon Capt, *Study in Pyramidology*, p. 150; Joseph P. Farrell, *The Giza Death Star*, p. 263.

I further noted that

- The granite **(and limestone)** stone courses function as the coil windings of the system, their resistance multiplying the voltage of the impulse output;
- The granite **(and limestone)** stone blocks function as the segments of the system, multiplying the voltage of the impulse output.[181]

The angle of the faces of the Pyramid, 51° 51' is precisely that of a quartz crystal, which means also that in addition to being a waveguide, the Pyramid itself is designed to be a gigantic crystal, with the stones of the structure itself functioning as the lattice nodes of the crystal. But how does one know that the purpose of the pyramidal shape itself, and of the stone courses, is to function as windings within the coil of a system designed to make a coil function as a waveguide?

5. The Longitudinal Wave Form in the Stone Courses

The answer to that question occurs in a little-noticed and even-less-discussed diagram in Pyramid-as-prophecy researcher Peter Lemesurier's *The Great Pyramid Decoded.* The diagram shows the stone courses's respective thicknesses which, when graphed, exhibit [a deteriorating wave form with 26 distinct peaks. His commentary is reproduced here in full:

> Allowing for (the degree of error in Petrie's measurements), it will be seen from the graph that a number of surprisingly regular 'curves' results, interspersed with a number of sudden 'peaks'—features which may conceivably have an objective, exterior *in some field of specialist inquiry yet to be identified.* Internally, however, it is noteworthy that factorization of the course numbers of the 26 unmistakably peak courses produces as factors the numbers 1,2,3,4,5,6,7,8,9,10, 11, 12, 13, 19,29, 41,

---

[181] Joseph P. Farrell, *The Giza Death Star*, p. 264.

and 67 (all of them features of the Pyramid's geometric and/or arithmetical codes ...plus the further numbers 17, 23, 37, and 59. Since it would presumably have been perfectly possible for the designer to have chosen exclusively **prime** numbers for his peak courses—or numbers divisible by 2,3, and 5, say—it seems reasonable to see in his choice of peak courses a deliberate indication of the essential signals comprising his internal code and a hint that factorization might play a role in its application.[182]

But here, for once, the actual graphed physical data is more important than the numbers, for the following waveform results when one looks at the thickness of the stone courses:

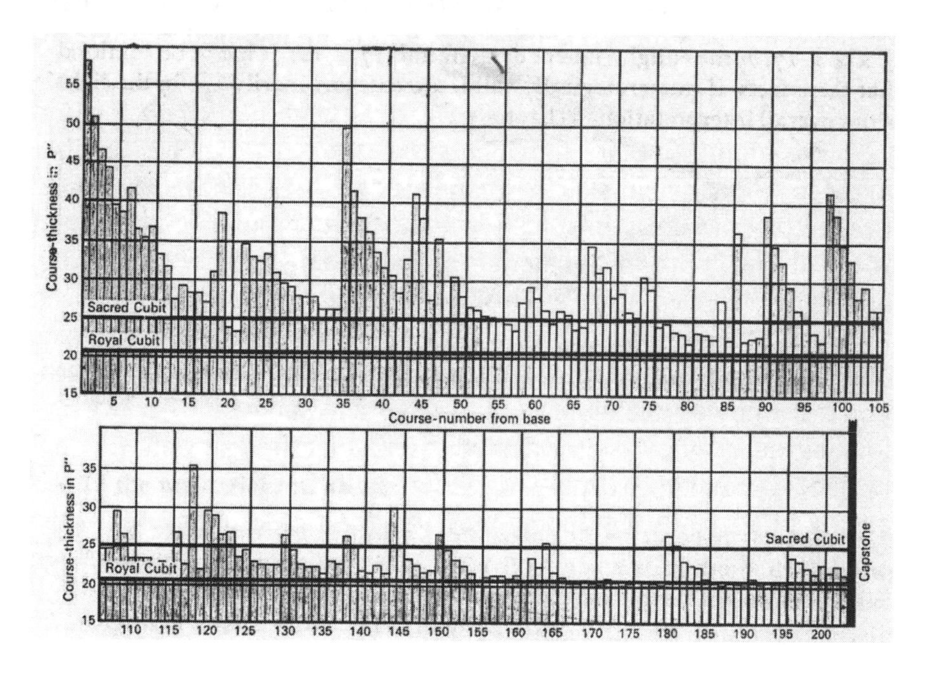

Lemesurier's Graph of the Thickness of the Stone Courses from the bottom of the Pyramid (upper graph, left) to the top of the Pyramid (lower graph, right)

[182] Peter Lemesurier, *The Great Pyramid Decoded* (New York: Avon Books, 1977, ISBN 0-380-43034-7), pp. 333-334.

Looking at this graph is illustrative and quite suggestive, for *a clear wave form is present* and if one takes *thickness* as a function of *frequency*, then one has a waveform that increases from low to high frequency. In other words, if one assumes a pulse beginning at the bottom of the structure, and moving over its surfaces to the top, the greater segmentation at the bottom of the structure increases the energy of the pulse, which is increasingly stepped *up* as it ascends in the structure, since the energy is confined to *an ever-decreasing surface area, increasing the scalar potential at any given point, and also since the frequency is constantly being increased.*

Our survey of the weapon hypothesis' essential hard components is now concluded. It is to be noted that, like all versions of the machine hypothesis, the weapon hypothesis acknowledges that the current structure at Giza is *incomplete*, and that accordingly we are looking at the *shell* of a machine, rather than the whole machine. The weakness of the hypothesis obviously lies in the fact that its putative missing components rely upon speculations about metamaterial crystals being gravity resonators, oscillators, and amplifiers. However, lest the reader have missed the point, this metamaterial crystals speculation itself emerges *from the wider context* not only of electro-acoustic or electro-gravitic impulse technology and from the analysis that we are looking at a very sophisticated piezo-electric version of that technology, but also from the wider context of *textual* evidence clearly suggesting a weapon function for the structure, the *only* structure which in its dimensional analogues reaches unto heaven itself. Because it is an *impulse* technology it is principally and always a longitudinal wave technology, a *gravity* oscillator and amplifier. Once this has been said, a weapon function would have been immediately apparent to its designer(s) and builder(s).

8
A BRIEF LOOK AT THE OTHER PYRAMIDS

"By applying Tesla's technology in the Great Pyramid, using alternating timed pulses... we may be able to set into motion 5,273,834 tons of stone! If we have trouble getting the Great Pyramid going, there are three small pyramids nearby that we can start first to get things going."
Christopher Dunn[1]

"(Bob Vawter) found overtones and resonance erects that recorded at different frequencies than those measured previously by other researchers in the Great Pyramid. This evidence enable us to speculate that the Per-Neters may have been 'tuned' to different frequencies to resonate harmonically with each other."
Stephen S. Mehler[2]

"On investigating the two pyramids at Dashur, I was struck by the fact that they both have exactly the same height, 105 meters, and the same final slope in their upper parts of 43°22'. This cannot have occurred by coincidence."
Alan F. Alford[3]

ANY SURVEY OF THE WEAPON HYPOTHESIS must also give some account of the other pyramids in Egypt and, for that matter, the world. Clearly, no other pyramid has the distinction that the Great Pyramid appears to have in the ancient texts with which we began this book; both in those texts and to any

---

[1] Christopher Dunn, *The Giza Powerplant*, p. 149.

[2] Stephen S. Mehler, *The Land of Osiris* (Kempton, Illinois: Adventures Unlimited Press, 2001), p. 120. Mehler uses the term "per-neter" as an ancient Khmetian term for "pyramid" Mehler is referring to the Red Pyramid at Dashur, some miles south of Giza.

[3] Alan F. Alford, *The Phoenix Solution: Secrets of a Lost Civilisation* (London: Hodder and Stoughton, 1998), p. 61.

unbiased examination of the structure, it appears as something *sui generis*, something so unique in the high standard of its perfection and the purposes for that perfection that the existence of other pyramidal structures in the world does not argue against that function and perfection, but rather, exhibits them all the more starkly.

Nevertheless, *some* account of them should be attempted, especially as the Giza compound and Egypt itself are home to more pyramids than just the Great Pyramid. The pyramids of Cephren and Menkaure at Giza are almost as famous as the Great Pyramid, as are the Red and Bent Pyramids some miles south of Giza at Dahshur. And of course, there are pyramidal temples in India, and the well-known stepped pyramids of Tikal and other locations in Meso-America, or the even more famous and mysterious pyramids of Teotihuacan near Mexico City, which were long since abandoned by whomever built them long before the Aztecs showed up and appropriated them into the empire of the Mexica. There clearly was some sort of global "pyramid culture," so how does one account for it and for the *sui generis* nature of the Great Pyramid at one and the same time, and in a manner that makes consistent sense?

It was partially in answer to that question that I wrote with my friend Dr. Scott deHart in my book *The Grid of the Gods*.[4] In that book, I laid out the thesis of Alan F. Alford on the dating of the Giza compound, and expanded upon it, due to its importance for the question of the relationship of other pyramidal structures to the weapon hypothesis. Accordingly, that component of *Grid of the Gods* is reprised here, along with the penultimate chapter of *The Giza Death Star Deployed*, in order to summarize my thinking

---

[4] Joseph P. Farrell with Scott D. deHart, *The Grid of the Gods: The Aftermath of the Cosmic War and the Physics of the Pyramid Peoples* (Kempton, Illinois: Adventures Unlimited Press, 2011, ISBN 978-1-935487-39-5). When we wrote that book, little did Dr. deHart or I suspect that within a few years our chapter on alchemy, Gothic cathedrals, and Notre Dame de Paris would be literally gutted by a kind of alchemical fire that transformed the interior of the famous cathedral by destroying it.

in other books on the relationship of other pyramidal structures to the Great Pyramid. As elsewhere in this book, these excerpts are cited verbatim, in sections beginning "[" and ending "]" with brackets, with additional or explanatory commentary that has been added to those sections appearing in boldface parenthetical comments within the main text.

When one looks at the Great Pyramid, and the Giza compound as a whole, in relationship to the other pyramidal structures on the globe, one comes to an intriguing realization:

> Giza lies at the center of the machine that was the world grid, for as we have observed, the ancient world prime meridian, and the placement of so many sites on the world Grid, are oriented with Giza being the prime meridian, and more specifically, with the prime meridian running through what would have been the apex of the Great Pyramid. Giza is, so to speak, the transmission gears at the center of the "machine" of the Grid, a great alchemical machine to manipulate the Grid itself.
>
> But Giza and its pyramids are more than just gears in a machine; it is also an alchemical working of a very different sort, for they have exercised a transforming fixation on the human imagination itself.[5]

Just how and why Giza should be at the center of a vast global machine brings us face to face with Alan Alford's complex hypothesis about the chronological levels of construction evident at Giza and Egypt at large.

A. Alford on the Dating and Design of Giza, and the Chronology Problem

[The problem of dating of the Giza compound is complex, and recently became much more so with the revelation by Dr. Robert Schoch that the water-erosion on the Sphinx meant that it was far older than dynastic Egypt, requiring a date of between

---

[5] Joseph P. Farrell, with Scott D. deHart, *The Grid of the Gods*, p. 293.

5000-7000 BC. But this, according to alternative researcher Alan Alford, implied that the whole question of dating the Giza compound itself had to be re-thought, for the compound itself as a whole was laid out according to an intricate geometrical plan **(as we shall see later in this chapter)**, a plan which included the Great Pyramid, and that

> Encompassed the Sphinx, its temples, the causeway and Khafre's pyramid, for it would seem that the position of the two Sphinx temples was determined by two intersection lines drawn from both of the two giant pyramids. Indeed, when we add to these relationships the common use of megalithic-style masonry in the temples of both (sic.) Sphinx and pyramids, it is easy to see why Egyptologists view all the structures of Giza as intimately linked, and thus roughly contemporary. The important implication of this is that one reliable dating has the potential to date all structures on the Giza plateau, hence the redating of the Sphinx is not an isolated issue, but has fundamental implications for our understanding of Egyptian history, and particularly to so-called 'Pyramid Age.'[6]

Close comparison of the construction quality evident in the structures at Giza revealed to Alford that there are at least three levels, or periods, involved, and it is worth citing what I remarked about Alford's conclusions in *The Giza Death Star Deployed:*

> The Sphinx, the temples, and the two giant pyramids at Giza were already present at the beginning of the Fourth Dynasty, and (the pharaohs) Khufu and Khafre simply adoptred and refurbished them, accounting for the radiocarbon dating anomalies. The society that designed and built the structures disappeared long before the Egyptians occupied them, *with an intervening period where the site was maintained by a small and elite priesthood.* Thus, in Alford's scenario, there are three distinct levels of the cultural occupation of Giza:

---

[6] Alan Alford, *The Phoenix Solution*, p. 6, also cited in *The Giza Death Star Deployed*, p. 27.

- The first level, responsible for the original construction of the major structures…
- The second level, a "remnant" of elite priesthood left behind at, or that came to occupy, the site…
- The third level, the Egyptian civilization itself.[7]

Thus, adopting Alford's conclusions and modifying them somewhat, I came to the conclusion that there were three levels, represented by three increasingly *declining* levels of construction perfection evident at Giza:

1) The oldest level, comprising the Great Pyramid itself, antedating ca. 10,000:
   a) since the re-dated Sphinx belongs to the *second* "less perfect" level of construction at Giza;
   b) since the Sphinx had been re-dated by Dr. Robert Schoch, based on its water erosion, to ca 5000-7000 BC; and,
   c) since there are ancient traditions that record the fact that when the Great Pyramid did have its casing stones on it, that a water mark was visible halfway up the structure, indicating that it ante-cated the agreed-upon date among alternative researchers for the Flood, ca. 10,000 BC;
2) The second, younger, but still pre-Egyptian level, lying somewhere between 10,000 BC and the Fourth Dynasty millennia later, represented by the Second Pyramid, the Sphinx, the various Sphinx temples, and possibly the third large Pyramid, Menkaure;[8] and,

---

[7] Joseph P. Farrell, *The Giza Death Star Deployed*, pp. 29-30.

[8] Alford himself notes the *sui generis* character of the Great Pyramid, and the organization of three levels of construction on that basis: "The character of the two pyramids is also very different. Whilst the builders of the Second Pyramid indulged in the occasional stylistic feature, such as a single layer of red granite casing stones at the pyramid's base, the builders of the Great Pyramid seem to have been concerned only with accuracy and stark functionality. Whilst the Second Pyramid might

3) The final, youngest, and purely Egyptian level, represented by the remainder of the structures at Giza, the causeways and the six smaller pyramids and also possibly by the third large Pyramid, Menkaure.[9]

But now the problem of dating grows more acute, for if one dates the first and second levels of Giza to antedate Egypt itself, the problem becomes one of *fixing* Giza within the wider context and implied technologies evident at such sites as Puma Punkhu, for one and the same technological skill is implied at both sites, and in both chronological levels of construction.][10]

Once one admits Alford's template of three levels of *declining construction techniques and perfection*, beginning with the oldest and most perfect, to the more recent and most declined, then the levels can be seen not only at Giza, but throughout the world. Once other structures in the world are looked at with Alford's template, the following items and locations seem to emerge:

1) the oldest and highest standard, ca. 10,000 BC or older, represented by structures such as the Great Pyramid, or sites such as Pumu Punkhu and Titicaca in Bolivia;

---

passably be a tomb or cenotaph, the Great Pyramid has a complexity akin to some giant machine which is beyond our comprehension.

"If preconceptions are put to one side and archaeology alone is our guide, the differences in design quality and aesthetics lead us inevitably to the conclusion that the two pyramids bear the fingerprints of *two separate pre-dynastic cultures at Giza.* One of these cultures, the designers of the Great Pyramid, seems to have been far in advance of anything else ever seen in Egypt. The other, the designers of the Second Pyramid, and perhaps also the megalithic temples and Sphinx, seems to have been equivalent in technical ability to Sneferu's deisgners at Dahshur." (Alford, *The Phoenix Solution*, p. 156, emphasis Alford's).

[9] Joseph P. Farrell, *The Giza Death Star Deployed,* pp. 31-32.

[10] Joseph P. Farrell, with Scott D. deHart, *The Grid of the Gods*, pp. 297-300.

2) an interim layer, representing pre-classical and pre-dynastic civilizations ca. 80000-5000 BC, represented by structures such as the Second Pyramid (Cephren), The Sphinx, the Valley temples, possibly the Osirion, and places such as Teotihuacan in Mexico or perhaps Mohenjo Daro in India;
3) the youngest level, from ca. 5000 BC into dynastic and classical times, represented by structures and sites such as Tikal, the third Pyramid of Menkaure, Dahshur with its Red and Bent Pyramids, and so on.[11]

1. Alford's Three Chronological Layers of Construction, and the Kings' Lists of Sumer and Egypt

If one admits all this, however, and if one also admits that these other famous sites around the world are laid out in a "grid" utilizing Giza and the Great Pyramid's apex as the prime merdian, then this suggests another problem, that "there was a continuity of ideology among those building the various structures around the world"[12] a continuity which in turn implies either a dedicated elite overseeing such projects, or an elite with very long (and hidden) life spans, or both.

With respect to these two possibilities of an elite – of an elite dedicated to a long construction project handed down from generation to generation, or even an elite some of whose membership possessed long and concealed lifespans overseeing such a long and global construction project—and the three chronological layers of construction, it's worth noting that both the Sumerian Kings' list as recounted by Berossus, and the Egyptian Kings' list as recounted by Manetho both mention *three* layers of kings:

1) the oldest layer, being the "divine" kings, or gods, that descended from heaven and ruled over men on earth, and

[11] Q.v. Joseph P. Farrell with Dcott D. deHart, *The Grid of the Gods*, pp. 301-302.

[12] Ibid., p. 302.

possessing tremendously long life spans in the tens or hundreds of thousands of years;

2) a second layer, being the semi-divine kings, the "demi-gods" who were partly divine, and partly human, still possessing long life spans but shorter than those of the "god-kings"; and finally,
3) the most recent layers, being the fully human kings, with the shortest life-spans.

While a comparison of the actual length of reigns from Berossus' and Manetho's Sumerian and Egyptian kings' lists with the dates for Alford's three construction layers do not dovetail, if one removes the dates from both lists, there is an amazing *correspondence of conception*: the oldest layer of construction in Alford's template is that of the highest standards and perfection, and the oldest layer of kings in both kings lists is the "divine" layer of "god kings," who could be expected to execute such intensely perfect constructions. The demi-god kings, part god and part human, would have a declined but still high standard of execution, and the final layer, the merely human, would have the lowest standard. When one adds the long life spas of the divine kings, the slightly shorter life-spans of the demi-god kings, the *reason* for the standard of perfection and its slight decline over time becomes evident.

To put all this as succinctly as possible, *the three layers of kings in the Sumerian and Egyptian kings' list rationalizes and dovetails almost perfectly with the three layers and standards of construction in Alford's template.*

B. Rotating Giza:
1. The Twists in the Two Large Pyramids: Torsion and "Rifling" the Waveguides

With regards to the plan of the compound at Giza exhibiting an intentionality consistent over millennia, mention must be made of two features that indicate that the structures and the compound itself were meant to *rotate* energies. In the previous chapter we made the case, based on Tesla's and T.T. Brown's electrical impulse technologies, that the Great Pyramid functioned both as a kind of piezo-electric "analogue electrical coil," but even more importantly, as a segmented *waveguide* for a longitudinal electrical pulse in the medium.

But one must not think of this pulse in a "straight-line" linear sort of way, but also as a *twist* or, to give it the more technical term, a *torsion* in the lattice of space-time. A simple analogy, and one which I have used many times to explain the effect of torsion on that lattice, is that of taking an empty soda can, representing space-time's lattice, and then wringing it like a dishrag between one's two hands. The resulting twists, wrinkles, and compression of the can represents the torsion itself.

[With these thoughts in mind, we now take a closer look at the two large pyramids of Giza, for there we find the idea of torsion, of a *twist*, present in the structures themselves. It was the renowned Italian metrologist Stechhini who observed that the Great Pyramid's four sides were all of slightly different lengths and angles from each other, producing a slight twist to the northwest corner of the structure.[13] Many who have studied the Great Pyramid have commented on this feature, calling it a "slight imperfection" and other such phrases **(in an otherwise nearly perfectly aligned and measured edifice).**

But is it?

Given all the other extraordinary perfection in the structure, we must ask whether this is a likely explanation, or if this twist in the structure is there *intentionally*?][14]

As I noted in *Grid of the Gods* there is a strong argument that this twist in the Great Pyramid is fully intentional, because Sir Flinders Petrie, in his monumental and magisterial study *The Pyramids and Temples of Gizeh* observed [a number of curious features about the *Second* Pyramid. Observing that the top of the Second Pyramid still has some of its casing stones on it, and taking

---

[13] John Michell, *The New View over Atlantis* (New York: Thames and Hudson, 2001, ISBN 0-500-27312-x), p. 147.

[14] Joseph P. Farrell with Scott D. deHart, *The Grid of the Gods*, pp. 305-306.

theodolite readings on the angles, Petrie stated that "From this it is seen that the builders skewed round the planes of the casing as they went upward; the twist being +1'40" on the mean of the sides; so that it is absolutely—3'50" from true orientation at the upper part."[15] Additionally, unlike the Great Pyramid, whose alignment to the four cardinal compass points are almost perfect, the Second Pyramid is twisted even further off alignment to the true compass points; in short, the structure itself is twisted, and in addition, twisted off center relative to the compass points.[16]][17]

If one understands *both* pyramids as waveguides in an impulse technology, then the purpose of imparting a slight twist to the structures becomes apparent, for this twist or torsion will be imparted to the impulse as it moves over the surfaces of the structure, creating a torsion in the medium itself as the longitudinal impulse moves through it. It is analogous to rifling a gun barrel or a cannon barrel: the rifling rotates the projectile when the gun is fired, and the rifling increases the energy, accuracy, and range of the projectile, or to put a finer point on it, rifling increases the efficiency of the cannon.

Additionally, if both pyramids are components in a large impulse machine technology, then the purpose for *two* such waveguides becomes clear, as an interferometric wavepattern can be created to steer the resulting impulses with great precision. By manipulating the *phase* of the impulses from each structure, further ability is obtained for targeting specific points either on the planet or in local space.

There is, however, [a *third* consideration that points, once again, very deliberately to the idea that the entire Giza compound was designed *as an analogue of torsion*, **(this time not in the individual structures but rather in their location in the overall compound itself).** In *The Giza Death Star Deployed* I pointed out the little-known work of Howard Middleton-Jones and James

---

[15] W.M. Flinders Petrie, *The Pyramids and Temples of Gizeh* (Elibron Classics, 2007), p. 97.

[16]Ibid., p. 99.

[17] Joseph P. Farrell with Scott D. deHart, *The Grid of the Gods*, p. 306.

Michael Wilkie, *Giza Genesis: The Best Kept Secrets*. Whatever else one makes of the claims in their book, they did make an important discovery for any machine hypothesis of the Great Pyramid or for the overall Giza compound, and an even more important discovery for the weapon hypothesis.

2. Middleton-Jones:
The Virtual Rotation of The Giza Compound Itself, and "Tetrahedral Physics"

Their discovery was simple. If one looks down on the Giza compound using an imaginary line through the apex of the Great Pyramid, and using that line as an axis of rotation for the whole compound, an amazing thing happens. Here is the compound itself, with the three main pyramids, Khufu, Cephren, and Menkaure, and the various "causways," with the Great Pyramid at the lower right:

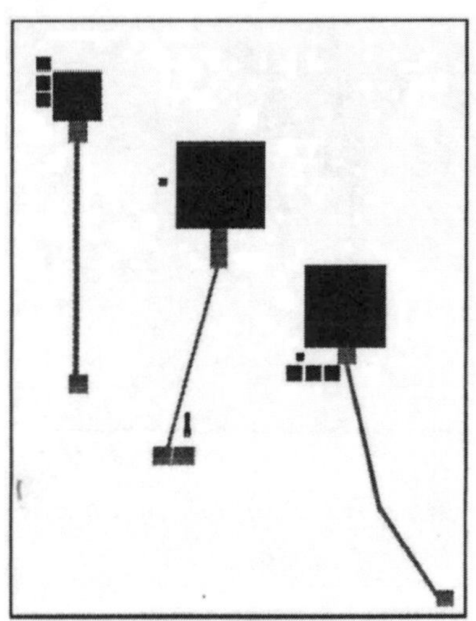

Middleton-Jones' and Wilkie's Diagram of Giza[18]

Now, rotate the entire compound through 120 degrees, and overlay the result with the original, and one gets this:

[18] Howard Middleton-Jones and James Michael Wilkie, *Giza-Genesis: The Best Kept Secrets,* Vol. 1 (Tempe, Arizona: Dandelion Books, No Date, ISBN 18933021408), p. 208.

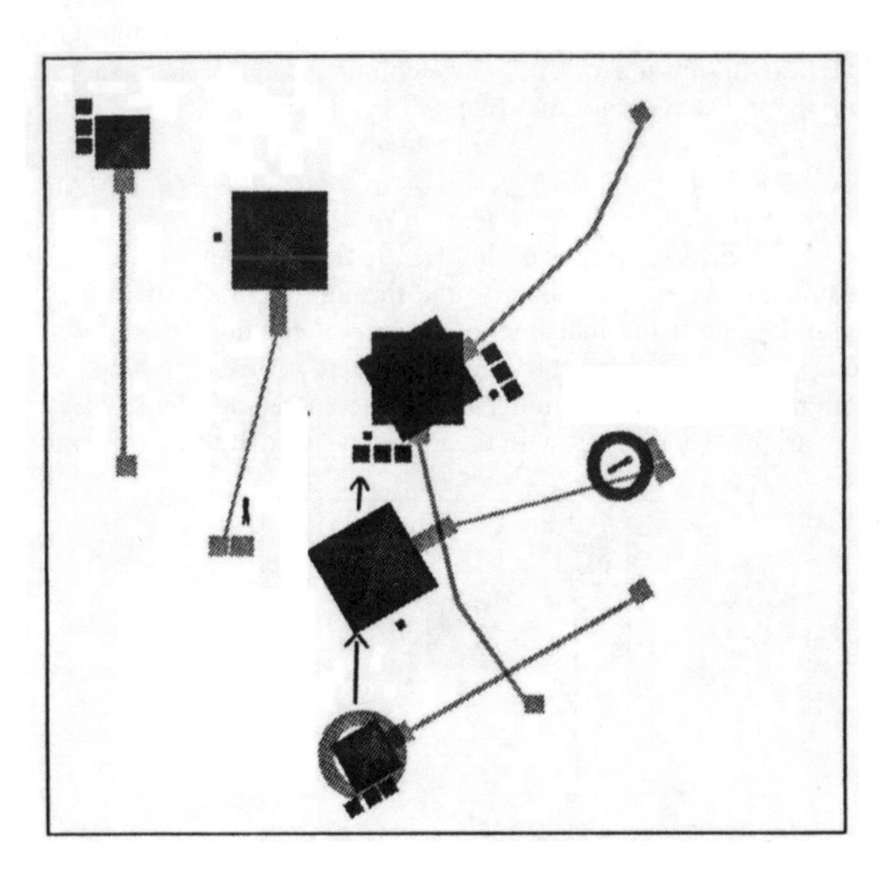

Middleton-Jones' and Wilkie's Diagram of Giza Rotated 120 Degrees[19]

Rotating the compound a further 120 degrees, for a total of 240 degrees, and overlaying *that* result with that of the previous rotation and the starting point, one obtains this:

[19] Middleton-Jones and Wilkie, *Giza-Genesis: The Best Kept Secrets*, p. 211.

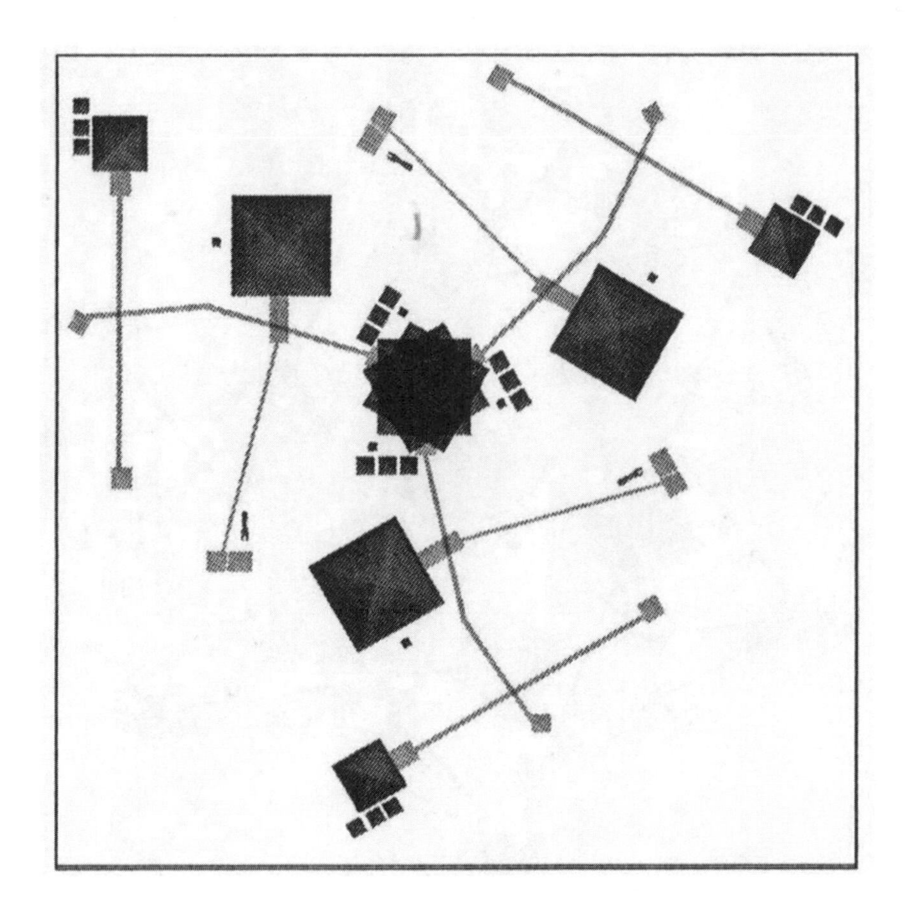

Middleton-Jones' and Wilkie's Giza Rotation through 240 Degrees[20]

And now, if one rotates this result through *60* degrees, and overlays it with the above result, one obtains this very well-known figure:

[20] Middleton-Jones and Wilkie, *Giza-Genesis: The Best Kept Secrets*, p. 211

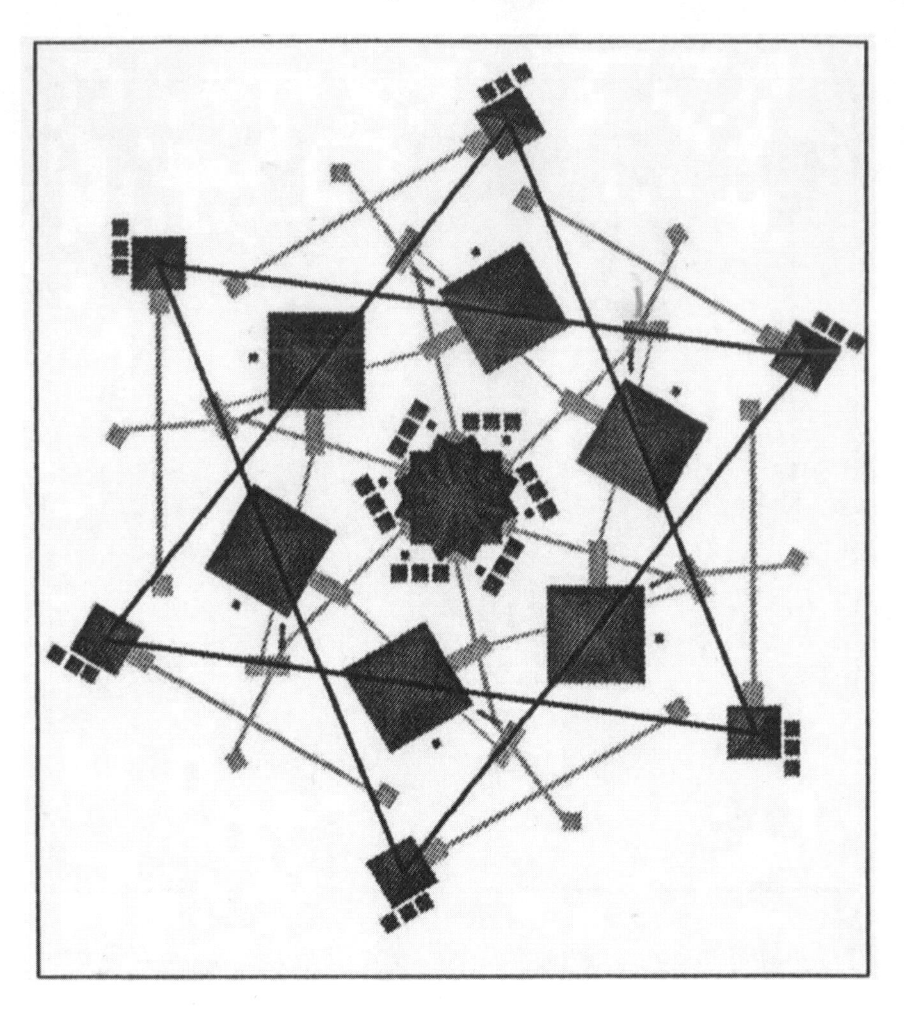

Middleton-Jones' and Wilkie's Rotation of the Giza Compound through six rotations of 60 degrees each[21]

Why would a two-dimensional surface of Giza with permanently placed buildings be said to rotate?

Leaving aside Middleton-Jones' and Wilkies' book, and continuing on our analysis that the Great Pyramid was designed as an electrical impulse technology of extreme power and

[21] Middleton-Jones and Wilkie, *Giza-Genesis: The Best Kept Secrets*, p. 213.

sophistication, *such rotation of the compound could be accomplished acoustically and/or electro-acoustically with the three other main pyramids.*

What would the point of this be?

The compound is a two-dimensional analogue of a three dimensional object whose technical name is "two orthorotated spherically circuminscribed tetradedra," or the symbol one gets when one embeds a tetradehdron inside a sphere with the vertices touching the sphere's surface, and one vertex on the sphere's axis of rotation. This, as Richard C. Hoagland has commented on many times over the years, means that the other three vertices of the inscribed tetrahedron will fall at exactly 19.5 degrees north or south latitude on that sphere. If one embeds a second teatrahedren inside the same sphere, and "orthorotates" it, i.e., places it entirely perpendicular to the first tetrahedron, one obtains the same figure but now in three dimensions:

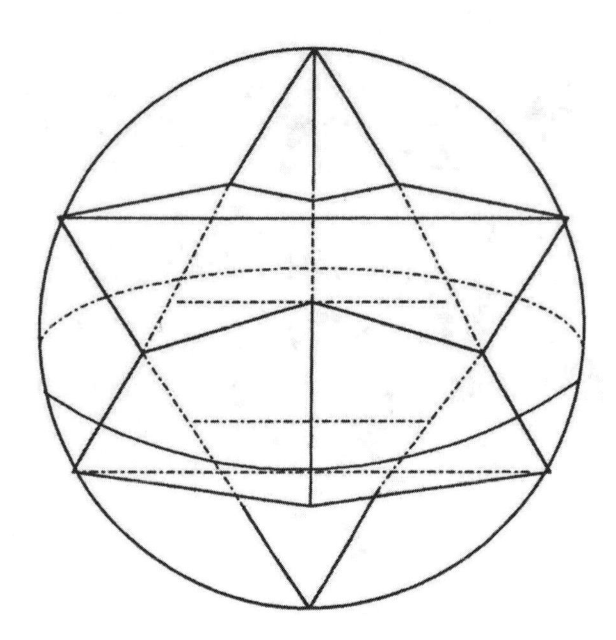

Two Spherically Circuminscribed Orthorotated Tetrahedra

Hoagland notes that many planets in the solar system contain upwellings of energy at more or less 19.5 north or south latitude on the planet, depending upon the orientation of the north magnetic pole and its axis of rotation. Thus, the great "red spot" or storm on Jupiter is at this approximate south latitude, the large volcano Olympus Mons on Mars is at this north latitude; on Earth volcanoes in Hawaii are at this approximate north latitude; on the Sun, sunspots occur at this approximate north *or* south latitude, and so on. Hoagland hypothesizes that this energetic upwelling is the result of any spinning spherical mass *transducing energy from the simple geometry of embedded tetrahedra as a hyper-dimensional characteristic of all such masses.* If Hoagland is correct, then the discovery of a *two dimensional* analogue of a compound that could itself be a two dimensional rotational analogue of a three-or-more dimensional object, then again, this fits the pattern we are accustomed by now to seeing: that every means possible was embodied and embedded there to make the *most efficient oscillators possible.* In this case it is also to be noted, once again, that the systems and energies *are planetary in scale.* The speculation, in fact, gives a more solid speculative and hyper-dimensional basis to the electro-acoustic nature of the weapon hypothesis, and moreover, to the science fiction speculation of Eando Binder's novel outlined in Chapter Three, Part One of the present work.

C. Alford and the Red and Bent Pyramids of Dahshur

In the standard Egyptological narrative, the stepped pyramid of Meidum, and then the two large pyramids at Dahshur—the Red Pyramid and the Bent Pyramid—came prior to the "perfection of the art" of pyramid building at Giza. This is, of course, to provide evidence for the narrative of "progress" and in part to root the Great Pyramid and the Second Pyramid at Giza firmly within dynastic Egypt, since there is abundant evidence that the pharaoh Sneferu really *did* build the structures at Dahshur. In this standard narrative, the Bent Pyramid is used to give evidence of a "mistake." According to this idea, Sneferu began the Bent Pyramid at the relatively high angle of 54°, whereas the Great Pyramid itself is inclined to only 51^{0}, as was seen previously.

Being unable to complete the building at this angle, Sneferu continued its construction at the easier angle of 43°. Being unsatisfied with *this* result, Sneferu supposedly built the Red Pyramid. After this false start of the decaying stepped pyramid at Meidum, and the "mistake" at Dahshur, the art gained its peak of perfection with the Great Pyramid.

So goes the Egyptology narrative.

But Alan Alford has a very different idea.

[Originally sheathed in brilliant white limestone casing like the Great Pyramid, the Red Pyramid of Dahshur gets its name from the red stones revealed when its casing stones were stripped away through the years. While the surface area that the Red Pyramid covers is comparable to that to the two giant pyramids of Giza, the Red Pyramid has a much smaller mass, given that its sides slope at an angle of 43° 22'.[22] This would make it an "obtuse" pyramid in relation to the side to height ratio of $\pi/2$. Like the Great Pyramid, it has an entrance leading down a sloping passage, which ends at a chamber with a corbelled roof of eleven corbels, sitting at roughly ground level, similar to the Great Pyramid's Grand Gallery in its corbelling, dissimilar in that it is neither inclined nor above ground level.[23] Another short passage leads to another similarly corbelled chamber directly beneath the pyramid's apex.[24]

Whatever its mysterious design features were meant to accomplish, it most certainly was never a tomb, which may be confirmed by examining the engineering "mistake" of the Bent Pyramid, which, after the Great Pyramid itself, may be the "most interesting pyramid in the whole of ancient Egypt."[25]

The more southern of the two Dahshur Pyramids, the Bent Pyramid rises to exactly the same height as the Red Pyramid, approximately 344.48 feet. Its upper angle of inclination, 43° 22' is exactly the same as the angle of inclination of the Red Pyramid. For the first lower third, however, its lower angle of inclination is 54° 28'. And, unique to all the Egyptian pyramids in this duality,

---

[22] Alan F. Alford, *The Phoenix Solution*, p. 275.

[23] Ibid., p. 52.

[24] Ibid.

[25] Ibid., p. 55.

the duality seems to be an intentional design feature, since the Bent Pyramid has both northern *and western* entrances.[26]

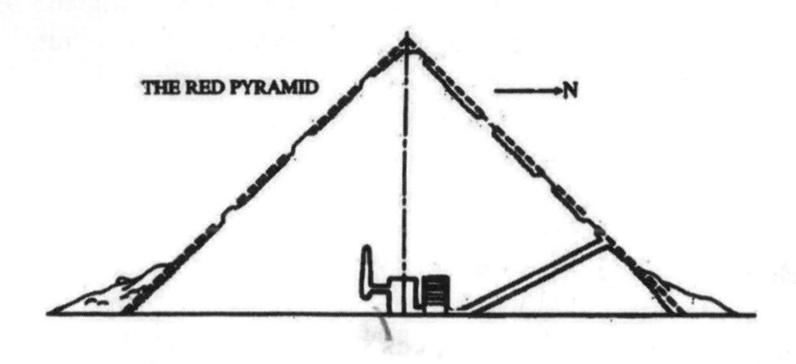

Alford's Diagram of the Red Pyramid of Dahshur[27]

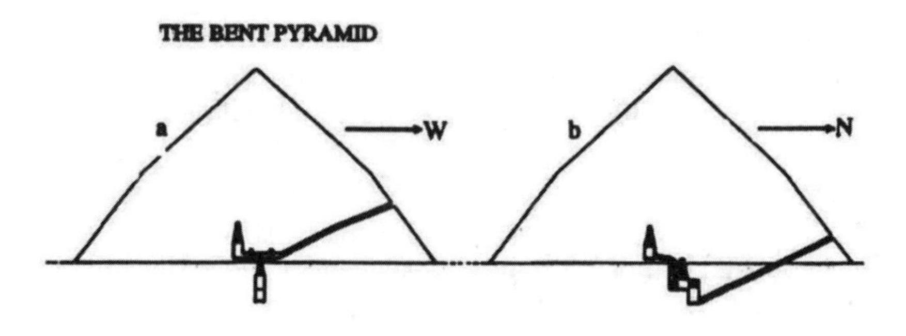

Alford's Diagram of the Bent Pyramid of Dahshur[28]

The Bent Pyramid's chambers, unlike the Red Pyramid whose chamber lies above ground, lie partially above, and partially below ground. Duality again. The passages themselves are curious studies in duality. The northern passage descends at a little more than 28 degrees **angle** but changes approximately half way down to a little more than 26 degrees, and the western passage,

---

[26] Alan F. Alford, *The Phoenix Solution* p. 55.
[27] Ibid., p. 52,
[28] Ibid., p. 55.

descending at 30 degrees initially changes in mid-course to a little more than 24 degrees.[29] At the end of this, two "portcullis" systems of limestone slabs are found, which may have had the function of :sonic baffles" similar to Dunn's hypothesis for the Antechamber in the Great Pyramid.

Alford then remarks that when he first visited Dahshur, he was conditioned by the standard view of Egyptology to find these two Pyramids examples of the inferior workmanship of the Egyptians, who were still learning the art of pyramid building prior to Giza.

> I had been seduced into believing that the builders of the Bent Pyramid were over-enthusiastic amateurs who had lost confidence in their design, in contrast to the professionals who had built the near-perfect pyramids of Giza. However, having now visited Dahshur, and seen with my own eyes the remarkable quality of *the Red Pyramid*, I am forced to question the conventional wisdom, and ask whether the Bent Pyramid was of an equivalent high standard. One man who certainly seemed to think so was the father of modern Egyptology, Sir Flinders Petrie, who inspected both of these structures in the 19th century. Petrie stated that "the general work of this pyramid is about equal to that of the Larger Pyramid of Dahshur." He also observed that the exterior casing of the Bent Pyramid had "good and close joints" and was "of the same quality as that of the Second Pyramid of Gizeh."[30]

So how *does* one explain the apparent "design failure" of the Bent Pyramid and ts curious dualities?

By geometry, of course.

Under geometric analysis, the design of the Bent Pyramid can hardly be qualified as a failure, but rather as part of a flawlessly executed deliberate plan and design. For one thing, the fact that the Bent Pyramid is of exactly the same height as the Red Pyramid is hardly coincidental.[31]

---

[29] Alan F. Alford, *The Phoenix Solution*, p. 56.
[30] Ibid., p. 60.
[31] Alan F. Alford, *The Phoenix Solution*, p. 61.

> Furthermore, those who know their pyramid geometry will understand that the slope of 43° 22' is one of two significant angles which are a whole number ('N') function of pi, where the following formula is operative:
>
> H=N x S/2pi, where H—height of the pyramid and S = length of a side in its base.
>
> The logic here is very simple. "n" represents the number which determines the slope of the pyramid. If N is exactly '4' the pyramid will have a slope of 51°50' as in the Great Pyramid of Giza. If N is exactly "3," on the other hand, the slope will be 43° 22', which is the final slope of both pyramids at Dahshur.[32]

But, asks Alford, "what would it mean if we could prove that the slope had been changed *at a significant height* in the Bent Pyramid?" Such a discovery would "certainly undermine the weight of that particular argument" of Egyptology that the change was due to a design flaw and the inability of Sneferu's unpractised engineers in building a pyramid at such a steep angle.[33]

As Alford explains,

> An *emergency* change should produce a non-significant number, not a whole number like '3' or '4' (as above). An *intended* change, on the other hand, would produce a significant number, of which we might only envisage two possibilities in the myriad total outcomes—the mid-point 3.5, or pi itself, 3.14. Let's try it:
>
> Bent Pyramid Height 105m = N x (side 188m/2pi)
> Thus N=105/29.92
> Thus N = 3.5[34]

But this "amazing result" is significant not only for the fact that the Bent Pyramid's design was hardly accidental, but **because it** also suggests that it, at least, was built *after* the Red Pyramid and the Great Pyramid, since the whole number N in the latter cases is 3

---

[32] Ibid.
[33] Ibid.
[34] Ibid., p. 62.

and 4 respectively. **(To project our hypothesis of acoustic resonance into this situation, the Bent Pyramid would appear to be designed to be resonant to both structures.)**

(All of this) suggests a difficulty for the standard chronology of Egyptology, which maintains that Sneferu built the Dahshur pyramids and then his son Khufu the Great Pyramid at Giza.

> It is this lack of precedent and evolutionary development, this sudden confidence and expertise of Sneferu, which is so troubling to me, and indeed to Egyptologists. These giant pyramid projects at Dahshur were planned and executed by experts. Of course they are within the limits of human achievement, but only given the necessary experimentation and learning experience. But where are the experiments, and where did this learning experience come from? It is almost as if Sneferu recaptured a lost knowledge, almost as if he acquired the keys to an ancient library of wisdom—a parallel to the modern idea of a lost 'Hall of Records."[35]

In other words, the Egyptologists' explanation of Giza being presaged by Dahshur, with Dahshur as the "learning experience" and Giza as the "perfection of the art," really explains nothing. It merely pushes the mystery back further, and further muddles the chronology. How did Sneferu's engineers build the two marvels of Dahshur? Where are the "experiments"?

The Bent Pyramid's incorporation of 3.5 also strongly suggests that the Great Pyramid was already in existence when Sneferu undertook to build the Red and Bent Pyramids, further confounding the standard chronology and buttressing Alford's claims for the antiquity of the Giza structures. In yet another Giza-Dahshur parallel, it is known that the Bent Pyramid was cased with the yellowish limestone that also cased the Second giant Pyramid of Giza, and that the Red Pyramid was once cased with the same pure white limestone as once cased the Great Pyramid. And that

---

[35] Alan F. Alford, *The Phoenix Solution*, p. 79.

begs the question of just which site was the original, and which was the imitator.[36]

Intrigued by this question, Alford took a ruler and connected the apex of the Red Pyramid with that of the Great Pyramid on a map, and drew another line connecting the apex of the Bent Pyramid with the Second Pyramid, and discovered that over a distance of 2 kilometers, the lines were exactly parallel. Since the two giant Giza Pyramids were built closely together on a clear diagonal plan, "it does not take a genius to see which pair of pyramids has been oriented to the other, and it is thus evident that Sneferu built his pyramids *after and not before* the Great Pyramids of Giza. Such a conclusion is entirely consistent with the archaeological evidence which reveals Giza to have been an important site in the 1st Dynasty, in contrast to Dahshur which had little importance prior to the reign of Sneferu."[37]

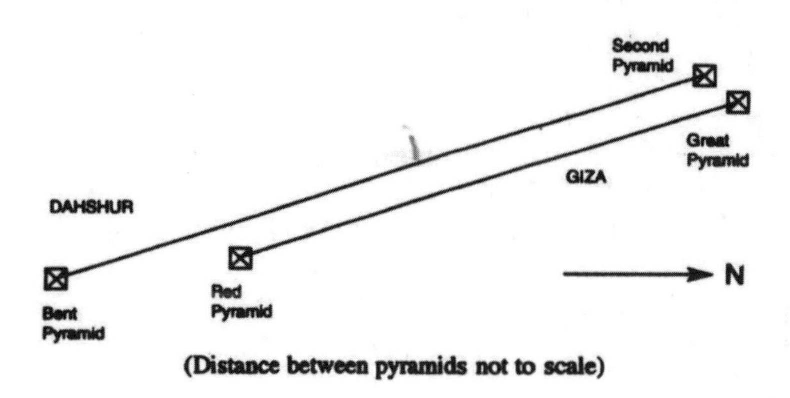

Alford's Diagram of the Giza-Dahshur Alignments of the Red and Bent Pyramids' Alignments to the Great Pyramid and Pyramid of Cephren at Giza[38]

---

[36] Alan F. Alford, *The Phoenix Solution*, p. 80.
[37] Ibid.
[38] Alan F. Alford, *The Phoenix Solution*, p. 81.

That Sneferu was responsible for the building of the two great Dahshur pyramids is beyond question. But what possibly could their function have been? If we accept Alford's idea that (Sneferu) indeed had access to, or rediscovered, some lost cache of knowledge, and tie it in with our own hypothesis and those of Dunn and Childress, then the function of the Dahshur monuments seems to suggest itself.

> Robert Vawter, trained as a musician and acoustical engineer as well as a field archaeologist, was able to do preliminary sound experiments and recordings in September 1997 that indicated the Red Pyramid creates harmonic resonance at a different frequency than other pyramids.... He found overtones and resonance effects that recorded at difference frequencies than those measured previously by other researchers in the Great Pyramid. This evidence enabled us to speculate that the (Pyramids) may have been "tuned" to different frequencies to resonate harmonically with each other.[39]

Moreover, as Mehler observes, the meaning of the name "Sneferu" itself may be a clue **(that the)** Bent Pyramid was not a mistake at all:

> The Bent Pyramid is a true (Pyramid) and was purposely built the way it was for principles of energy production through acoustical harmonic resonance by virtue of its unique shape. The Red Pyramid, at a 43 degree angle, may vibrate in a specific harmonic with the Bent Pyramid, and that also may be the reason for the term "Double Harmony" (Sneferu) at the site, not a specific king's name.... (Dunn) suggested that the Bent Pyramid, with its two angles of construction could produce multiple frequencies of sound, and this may be, in itself, the reason for the term Double Harmony.[40]

Given the geometric relationships between Giza and Dahshur, the evident design employed in the construction both of the Red

[39] Stephen J. Mehler, *The Land of Osiris,* pp. 70-71.
[40] Ibid., pp. 72-73.

Pyramid and of the Bent Pyramid, and the established resonant properties of the two pyramids of Dahshur, one must conclude that they, too, are coupled harmonic oscillators, *of the structures at Giza.* They may very well represent some sort of sophisticated power plant. But they are, for all that, the imitators of a much more formidable pattern and physics evident at Giza.][41]

What Alford's Giza-Dahshur hypothesis does, in the long run, is to underline, once again, and from a purely chronological point of view, that the Great Pyramid when compared to all the other pyramidal structures on the globe is an edifice *sui generis*, both in terms of the quality of its construction, the numerous dimensional analogues present in the structure, the numerous textual references referring to it, and of course its function.

Even after the removal of those hypothesized metamaterial components in the Grand Gallery which "made it work," it retained those more benign odd powers and abilities that researchers have come to discover about pyramids over the years, one of which is their uncanny ability to preserve organic materials without putrefaction. It should be no surprise, therefore, that the land with which they are most associated, Egypt, should eventually associate them with its whole death-and-immortal life mythology, nor that, a continent and ocean away in Mexico, they should come to be associated with sacrifice and immortality of a different kind.

As exemplified by the Great Pyramid, pyramids were a technology to manipulate the physics of the heavens themselves, and if the texts and its own structural analogues are any clue, it was able to do so with an unequalled power for destruction such that it was a threat even to the gods themselves, and had to be rendered permanently inoperable, with only the shell of the machine remaining at Giza as a stark reminder, near another monument, the Sphinx, which Arab tradition records as "the father of terrors," near an Egyptian city—Cairo—whose Arabic name, Al Kaireh means Mars, the god of war…

---

[41] Joseph P. Farrell, *The Giza Death Star Deployed,* pp. 274-280.

The original entrance to the Great Pyramid

The Giza Pyramids: Menkaure on the far left, Cephren in the middle, and on the far right Cheops or Khufu (The Great Pyramid)

The original entrance to the Great Pyramid, and the forced 9th century entrance of Caliph Al Maimoun on the lower right behind the gathering of people.

Appendix A: A Wider Context: The Platonic Turn, Stochastic Processes, Occult Symbols, and the Analogical universe

IN THE ORIGINAL *GIZA DEATH STAR* and *Giza Death Star Deployed*, I spent a great deal of time providing the physical analogues of what are normally viewed as occult practices, and seldom (if ever) viewed as simple analogues of the physics of stochastic processes, their self-organizing properties, and the underlying cosmology of analogies, or analogical cosmology, they point to. Additionally, that section of *The Giza Death Star Deployed* also pointed out the very strange and very "non-Egyptological" list of people who had shown an interest in the Great Pyramid. While not essential to the weapon hypothesis I believe this analysis is germane to it, and of its own intrinsic value, so I reprise here what I wrote in *The Giza Death Star Deployed* for the completeness of the record.

[...(This) mention of a "Hermetic Weapon" does deserve some further commentary. In ... *The Giza Death Star* I cited William Henry's *One Foot in Atlantis*, an excellent survey of the occult influences operating in the circles of Roosevelt, Churchill, and Hitler. There, I pointed out Henry's list of parallels between standard occult (or ceremonial magick) talismans of power, the Tarot deck, and a modern pack of playing cards:

Occult Symbol:	Cauldron	Sword	Spear	Stone/Crystal
Tarot:	Cups	Swords	Wands	Pentacles
Modern Cards:	Hearts	Spades	Clubs	Diamonds

A further parallel exists in the accoutrements of the practitioner of ceremonial magick:

Occult Symbol:	Cauldron	Sword	Spear	Stone/Crystal
Tarot:	Cups	Swords	Wands	Pentacles
Modern Cards:	Hearts	Spades	Clubs	Diamonds
Magick:	Chalice	Dagger	Wand	Crystal

Ralph Ellis, in his *Thoth: Architect of the Universe*, points out a possible "paleophysical meaning" behind the minor trumps of the Tarot and the modern pack of playing cards, stripping away one layer of a possible physics origin for something as simple as a deck of cards:

Item:	**Number:**	**Physics Analogy:**
Number of cards	52	Number of weeks in the year
Number of cards per Suits	13	Number of lunar months
Number of Picture Cards	12	Number of terrestrial months
Number of suits	4	Number of seasons
Number of spots or (pips)	364	Approximate number of days in the Earth's Solar orbit

Interestingly enough, the only survivor of the Major Trumps from the Tarot deck in the modern playing cards deck is The Fool, or as it is more commonly known, the "Joker". Giving that card an arbitrary value of 1.234 would increase the number of "pips" or spots in the playing card deck to 365.234, a number reflecting the current calendrical system of counting the number of days in a year. But one may go further, much further, in uncovering possible encoded deep layers of physical meanings latent in the now garbled traditions of esoteric practice. For example, in the typical "Tarot reading" the reader is supposed to clear his mind, concentrate on the question being asked of the cards, and then begin the "shuffle." What basis in physics might these seemingly simple acts have? Since quantum mechanics and more recent scientific investigations into the relationship of consciousness and physical reality have pointed out some sort of connection between Observer and observed effect, the following relationships suggest themselves:

Tarot Reading:	**Possible Underlying Physics**
Clearing the mind and focus-Ing on the question	Posited links between the mind, and the structured potential of the zero point energy or vacuum flux
The Deck itself	The Infinite information potential of the field
the Shuffle	Analogue random number generator
the Spread	Union of the previous steps, i.e., of form (ειδος) and matter (υλη); the structured potential of information in the field

Extending this analogical analysis even further, recontextualizing the four standard talismans of power reveals more tantalizing possibilities when viewed in the contexts of electromagnetics and topology:

Occult Symbol:	Cauldron	Sword	Spear	Stone/Crystal
Tarot:	Cups	Swords	Wands	Pentacles
Modern Cards:	Hearts	Spades	Clubs	Diamonds
Magick:	Chalice	Dagger	Wand	Crystal
Electromagnetism:	Capacitor	Waveguide	Antenna	Crystal/ Lattice
Topology:	Basins of Attraction	Bifurcation	Maps	Matrix/ Metric

Notice how this chart, ascending from the images of the occult, Tarot and modern playing cards symbolism to physics and topology roughly corresponds with Fr. Francis Copleston's diagram of the Platonic turn (περιαγωγη) from the material to the intelligible world.... Combining the above table with Copleston's chart of the Platonic turn is very revealing (q.v. the last page of this appendix).

In any case, the high weirdness of people and associations investigating the Giza Death Star takes an even stranger turn during the twentieth century's scientific investigations of the structure, for almost all of them, without exception, were led by people with close ties to military or space research agencies. In 1996, for example, an Egyptian team headed by Dr. Farouk El Baz was supposed to open Gantenbrink's door on live television.[1] The event failed to materialize after much media hype. Even stranger is the fact that Dr. Farouk El Baz is a planetary geophysicist who worked with NASA on the Apollo moon landings.[2] A planetary geophysicist studying the Great Pyramid?

Stranger than this is Dr. Zahi Hawass himself, the Egyptian government's "tsar of Giza." Hugh Lynn Cayce, son of the American psychic Edgar Cayce and director of Cayce's ARE Foundation, claimed to have mentored and funded Hawass's doctoral studies in Egyptology in the United States, a claim that Hawass denies vehemently.[3]

Weirder still was the rumor that a tunnel was being dug from Davidson's Chamber to Gantenbrink's Door. Unlike many rumors, its source was none other than "Thomas Danley, an acoustics engineer and NASA consultant for two space shuttle missions, who specializes in 'acoustic levitation' (raising objects through the use of sound and vibration."[4] Danley discovered that Caviglia's tunnel, made in the 19th century during Vyse's infamous expedition, had recently been extended thirty feet beyond its original end. Informing a disconcerted Egyptian inspector, the Egyptian and his boss, Dr. Zawi Hawass, claimed to know nothing

---

[1] Gantenbrink's door is a relatively modern discovery of a small "door" with two metal poles some way up the "air shafts" from the Queen's chamber in the Great Pyramid.

[2] Robert Bauval, *The Secret Chamber*, p. 378.

[3] Ibid., pp. xxxviii-xl. Hawass actually attended the University of Pennsylvania on a Fulbright scholarship.

[4] Linn Picknett and Clive Prince, *The Stargate Conspiracy*, p. 79.

about it.[5] Either the Egyptian authorities were telling the truth and someone else was doing secret tunneling of the Pyramid, or the Egyptian authorities were lying and a cover-up was under way. Two years later the Egyptian authorities admitted that more tunneling was under way.[6] And according to Picknett and Prince, reliable sources confirmed that three new chambers had been discovered around the King's Chamber.[7]

In the *Giza Death Star Deployed* I also pointed out that former Manhattan Project scientist and Nobel Laureate Dr. Luis Alvarez was also involved in investigating the Great Pyramid.[8] [After dismissing his own research project, nothing further seems to have happened until 1973, when SRI International funded an expedition under the leadership of **physicist** Dr. Lambert Dolphin Jr. to look for hidden chambers under the Sphinx. Dolphin himself claimed that the expedition was but a continuation of Alvarez's research five years previously.[9] Like the Alvarez expedition, SRI's is no less questionable, for SRI International was but the renamed Stanford Research Institute, a famous think tank with close ties to the US Department of Defense and the various American intelligence agencies.[10]

Dolphin himself, a physicist, has his own peculiar associations. Not only is he tied to the American military and intelligence communities via SRI International, but he is also a typically "evangelical Christian fundamentalist" who apparently at

---

[5] Ibid., p. 80.

[6] Picknett and Prince, *The Stargate Conspiracy*, p. 81.

[7] Joseph P. Farrell, *The Giza Death Star Deployed*, pp. 88-92.

[8] Alvarcz rcccived the Nobel Prize in physics for his role in solving the problems associated with an implosion fission device for the plutonium atom bomb in the World War II Manhattan project. I believe him to be the "Mr. Alvarez" that the German infrared fuse inventor, Dr. Heinz Schlicke, debriefed in the Pentagon after his capture aboard the U-234 which was carrying highly enriched uranium 235 to Japan. See my books *Reich of the Black Sun*, and *The Nazi International*, from Adventures Unlimited Press.

[9] Picknett and Prince, *The Stargate Conspiracy*, p. 85.

[10] Ibid., p. 84.

the same time he was leading the Giza expedition was also helping Jewish fundamentalists find the original foundations of the temple, for the purpose of rebuilding it![11] In 1977 Dolphin and SRI returned to Giza, gaining the financial support of (Edgar) Cayce's A..R.E. foundation in 1978.][12]

A COMPARISON OF THE PLATONIC TURN TO PHYSICS AND ESOTERIC SYMBOLS

Copleston's Schema of the Platonic Turn				*Esoteric / Physics Correspondences*
MENTAL STATE		**OBJECTS KNOWN**		
Scentific understanding (η επιστημη)	**Intellection** (η ϖοησις)	**Sources/Principles** (αι αρχαι)	**Invisibles** (τα αορατα)	**Topology:** Basins of Attraction, Mappings, Matrix/ metric, Bifurcation
Knowledge (η γϖωσις)	**Discernment** (η διανοια)	**Mathematicals** (τα μαθηματικα)	**Intelligibles** (τα νοητα)	
Opinion-glory (η δοξα)	**Belief, Faith** (η πιστις)	**Living Things** (ζωα, κ.τ.λ.)	**Visibles** (τα ορατα)	**Physics:** Capacitors, Antennae, Crystals/lattices Waveguide
	Shadows, Images (η εικασια)	**Icons** (εικονες)	**Perceivables** (δοξαστα)	**Modern Deck:** Hearts, Clubs, Diamonds, Spades **Tarot Deck:** Cups, Wands, Pentacles, Swords **Magick:** Chalices, Wands, Crystals, Daggers

[11] Picknett and Prince, *The Stargate Conspiracy*, p. 85. It should be noted that a rebuilt temple and re-institution of animal sacrifices is a crucial component in the evangelical pre-millenianlist dispensational timetable for the second coming of Christ.

[12] Joseph P. Farrell, *The Giza Death Star Deployed*, p. 92.

Appendix B: Radiocarbon Dating of the Mortar of the Great Pyramid

MUCH HAS BEEN MADE OF RADIOCARBON DATING of the mortar in between the stones of the Great Pyramid, but in the case of the Pyramid, this technique faces two significant problems. In Dunn's machine hypothesis the Pyramid was filled with a hydrogen gas and additionally the coffer in the King's Chamber functioned as the optical cavity for a maser. ON this view the hydrogen gas may have been in an endothermic plasma state, excited by the piezoelectric properties of the structure, and in my view possibly experiencing a plasma pinch effect due to the presence of the maser. The maser may have even provided an optical channel to focus and convey a highly energetic plasma.

In any case, it is possible that all of this would have meant the presence within the structure of radioactivity, which would throw off any carbon-dating of the mortar of the structure, and, since the internal chambers of the building are closer to the ground than to the top, would subject the mortar closer to the chambers to more radiation, and hence the mortar at the bottom of the structure would appear younger than that at the top. Perhaps in some odd way Herodotus' odd remark that the Pyramid was built "from the top down" is a distant reflection of this very odd circumstance.

One of the effects of the pyramid shape *itself*, without the presence of any interior radioactivity, is the apparent ability of the shape itself – through processes presently unknown – to alter the half-life properties of radioactive materials. Alan F. Alford is the only author I know of who has questioned the value of radiocarbon dating of the mortar of the structure for precisely this reason:

> Finally, there is another important aspect of the pyramid radiocarbon dating project which has so far been overlooked, yet must be scientifically assessed, namely the ability of *the pyramid shape* to produce observed effects which defy currently known

> laws of science. Such effects are well-attested and whatever their cause, suggest that the pyramid environment is *different* from the general environment in which the science of radiocarbon dating has become established. As a result of these environmental anomalies, the decay rate of C-14 in the pyramid mortar samples may have been distorted to varying degrees, depending upon their positions in the overall pyramid shape, and, moreover, it is possible this effect is more pronounced in the larger pyramids, such as the Great Pyramid and Second Pyramid of Giza.[1]

This is significant, because the most recent radiocarbon dating, as Alford himself notes, returned dates for the Great Pyramid that were *400-500 years older than dynastic Egypt itself.*

But if this reading is a *distorted* reading due to the presence of radioactivity at some point in its past, or due to the effect of the shape and size of the structure itself, or a combination of the two, and if the *degree* of such distortion due to such factors is unknown and hence incalculable, then the Great Pyramid may be far older than anyone imagined.

[1] Alan F. Alford, *The Phoenix Solution*, pp. 625-626.

Bibliographies
of the original *Giza Death Star* Trilogy with the Additional Books and Articles Referenced Here

Books Consulted for, or Cited in, *The Giza Death Star:*

Alfvén, Hannes. "Cosmology in the Plasma Universe," *Laser and Particle Beams*, Vol. 16, August 1988.

Alfvén, Hannes. "Cosmology in the Plasma Universe," *Laser and Particle Beams.* Cambridge University Press. 1983, pp. 389-398.

Alfven, Hannes, "Existence of Electromagnetic-Hydrodynamic Waves," *Nature,* No. 3805, Oct. 3, 1942, pp. 405-406.

Alfvén, Hannes. "On Hierarchical Cosmology," *Astrophysics and Space Science,* Vol. 89., pp. 313-324.

Alfvén, Hannes. *Cosmic Plasma*. D. Reidel. 1981.

Aspect, Alain; Grangier, Philippe; and Roger, Gérard. "Experimental Tests of Realistic Local Theories via Bell's Theorem," *Physical Review Letters,* Volume 42, Number 7, 17 August 1981, pp. 460-463.

Bacon, Sir Francis, Lord Verulam. *The Advancement of Learning and the New Atlantis*. Oxford University Press. 1966.

Barrow, John D., and Tippler, Frank J. *The Anthropic Cosmological Principle*. Oxford University Press. 1988.

Bearden, Tom, Lt. Col. (US Army, Ret.) *Gravitobiology*. Tesla Book Co.

Benavides, Rodolfo. *Dramatic Prophecies in the Great Pyramid.* Mexico City. Editores Mexicanos Unidos, S.A. 1974.

Brian, Alice, and Galde, Phyllis. *The Message of the Crystal Skull: From Atlantis to the New Age.* Llewellyn Publications. 1991.

Brian, William L. II. *Moongate: Suppressed Findings of the U.S. Space Program: The NASA-Military Cover-Up*. Future Science Research Publishing Co. 1982.

Capt, E. Raymond., M.A., A.I.A., F.S.A., Scot. *A Study in Myramidology*. Thousand Oaks, California. Artisan Sales. 1986.

Childress, David Hatcher. *The Technology of the Gods: the Incredible Sciences of the Ancients.* Kempton,, Illinois. Adventures Unlimited Press. 2000.

Childress, David Hatcher. *Vimana Aircraft of Ancient India and Atlantis.* Kempton, Illinois. Adventures Unlimited Press. 1999.

Copleston, Fredericj, S.J. *A History of Philosophy*. Vol. 1, Pt. 1, *Greece and Rome*. New York. Image Doubleday. 1962.

Deyo, Stan. *The Cosmic Conspiracy*. New Revised Edition. Kempton, Illinois. Adventures Unlimited Press. 1994

Dunn, Christopher. *The Giza Powerplant: Technologies of Ancient Egypt*. Santa Fe, New Mexico. Bear and Company. 1998.

Fox, Hal. *Cold Fusion Impact on the Enhanced Energy Age.* Fusion Information Cemter. 1992.

Gray, Julian T. *The Authorship and Message of the Great Pyramid*. Cincinnati. E. Steinman & Co. 1953.

Hancock, Graham, and Bauval, Robert. *The Message of the Sphinx: A Quest for the Hidden Legacy of Mankind*. New York. Three Rivers Press. 1996.

Hancock, Graham, and Bauval, Robert. *The Orion Mystery: Unlocking the Secrets of the Pyramids*. New York. Crown Publishers, Inc. 1994.

Hancock, Graham. *Fingerprints of the Gods.* New York. Three Rivers Press. 1995.

Henry, William. *One Foot in Atlantis: The Secret Occult History of World War II and its Impact on New Age Politics*. Anchorage, Alaska. Earthpulse. 1998.

Herbert, Nick. *Quantum Reality*. New York. Anchor Books. 1985.

Hoagland, Richard C. *The Monuments of Mars: A City on the Edge of Forever*. North Atlantic Books. 1992.

Ignatovich, V.K. "The Remarkable Capabilities of Recursive Relations." *American Journal of Physics.* Vol. 576, No.

10. Pp. 873-878.

Kaku, Michio, and Thompson, Jennifer. *Beyond Einstein: The Cosmic Quest for the Theory of the Universe.* New York. Anchord Books. 1995.

Kaku, Michio. *Hyperspace: A Scientific Odyssey through Parallel Universes, Time Warps, and the 10th Dimension.* Oxford University Press. 1994.

Kaku, Michio. *Introduction of Superstings and M-Theory.* 2nd Ed. Springer-Verlag, Inc. 1999.

Katznelson, Yitzhak. *An Introduction to Harmonic Analysis.* Dover. 1976.

LaViolette, Dr. Paul. *Beyond the Big Bang: Ancient Myths and the Science of Creation.* Park Street Press. 1995.

Lemesurier, Peter. *The Great Pyramid Decoded.* New York. Avon Books. 1977.

Lepre, J.P. *The Egyptian Pyramids: A Comprehensive Illustrated Reference.* 1990.

LaViolette, Paul A. "The U.S. Anti-Gravity Squadron," in Thomas Valone, *Electrogravitic Systems: Report on a New Propulsion Methodology.* 2nd Ed. Integrity research Institute Publishers. 1995.

Liu, Y. "Acoustics, An Unofficial Introduction." www.stemnet.nf.ca/~yliu/acoustics.html.

McClain, Ernst G. *The Pythageorean Plato: Prelude to the Song Itself.* Nicolas-Hays, Inc. 1984.

Pepper, David M. "Applications of Optical Phase Conjugation," *Scientific American.* Vol 254, No. 1, Jan. 1986, pp. 74-83.

Pepper, David M. "Nonlinear Optical Phase Conjugation," *Optical Engineering: The Journal of the Society of Photo-Optical Instrumentation Engineers*, Vol. 21, No. 2, March-April 1982, pp. 157-183.

Plato, *Critias*, in *Collected Dialogues of Plato*, ed. E. Hamilton and H. Cairns, trans A.E. Taylor. Princeton University Press. 1961.

Scott, trans. and ed. *Hermetica*. Shambala.

Sitchin, Zechariah. *The Lost Realms: Book IV of the Earth Chronicles*. New York. Avon Books. 1990.

Sitchin, Zechariah. *The Wars of Gods and Men.* New York. Avon Books. 1995.

Sitchin, Zechariah. *Genesis Revisited.* New York. Avon Books. 1990.

Vassilatos, Gerry. "The Farnsworth Factor: The Most Notably Forgotten Episode in 'Hot' Fusion History," *Borderlands*, Second Quarter. 1995.

Bassilatos, Gerry. *Secrets of Cold War Technology: Project HAARP and Beyond.* Bayside, California. Borderland Sciences. 1996.

Wilson, Colin, and Flem-Ath, rand. *The Atlantis Blueprint: Unlocking the Ancient Mysteries of a Long-Lost Civilization.* Delacorte. 2002.

"X, Commander." *The Philadelphia Experiment Chronicles: Exploring the Strange Case of Alfred Bielek and Dr. M.K. Jessup.* Abelard Publications. 1994.

Books Consulted for, or Cited in, *The Giza Death Star Deployed:*

-------. "Tesla's 'Teleforce' Defensive Beam against Air Attack," *New York Times*, Sept 22, 1940.

Alford, Alan F. *The Phoenix Solution: Secrets of a Lost Civilization.* London. Hodder and Stoughton New English Library. 1998.

Barton, George A., trans. *Miscellaneous Babylonian Texts.* New Haven. Yale University Press.

Bauval, Robert. *The Secret Chamber: The Quest for the Hall of Records*. London. Arrow Books. 2000.

Bearden, Lt. Col. T.E. (US Army, Ret.) "Historical Background of Scalar EM Weapons," *Analysis of Scalar/ Electromagnetic Technology*. Tesla Book Company, 1990. Pp. 11-25.

Bearden, Lt. Col. T.E. (US Army, Ret.)"Introduction and Progress Report," *Analysis of Scalar/Electromagnetic Technology*. Tesla Book Company, 1990.

Bearden, Lt. Col. T.E. (US Army, Ret.). "Maxwell's Original

Quaternion Theory was a Unified Field Theory of Electromagnetics and Gravitation," *Proceedings of the International Tesla society*, pp.. 6/24-6/68.

Bearden, Lt. Col. T.E. (US Army, Ret.). "Scalar Electromagnetics and Antigravity," *Analysis of Scalar/Electromagnetic Technology*. Tesla Book Company, 1990. Pp. 73-87.

Bearden, Lt. Col. T.E. (US Army, Ret.). "Some Characteristics of the Phase Conjugate Wave," *Analysis of Scalar/ Electromagnetic Technology*. Tesla Book Company, 1990. Pp. 89-92.

Bearden, Lt. Col. T.E. (US Army, Ret.). "Soviet Phase Conjugate Weapons," *Analysis of Scalar/Electromagnetic Technology*. Tesla Book Company, 1990. Pp. 35-46.

Bearden, Lt. Col. T.E. (US Army, Ret.). "The Western Scientific Community's Record on Unorthodox Science," *Analysis of Scalar/Electromagnetic Technology*. Tesla Book Company, 1990. Pp. 47-54.

Bearden, Lt. Col. T.E. (US Army, Ret.). "USSR: New Beam Energy Possible?" *Analysis of Scalar/Electromagnetic Technology*. Tesla Book Company, 1990. Pp. 31-33.

Bearden, Lt. Col. T.E. (US Army, Ret.). *Gravitobiology*. Tesla Book Co.

Bennett, Mary, and Percy, David S. *Dark Moon: Apollo and he Whistleblowers*. Kempton, Illinois: Adventures Unlimtied Press. 2001.

Bohm, David. *Wholeness and the Implicate Order.*

Carlotto, Dr. Mark J. *The Martian Enigmas: A Closer Look.* North Atlantic Books. 1997.

Cheney, Margaret. *Tesla: Man Out of Time*. Laurell-Dell. 1981.

Childress, David Hatcher. *Atlantis and the Power System of the Gods: Mercury Bortex Generators and the Power System of Atlantis*. Adventures Unlimited Press. 2002.

Childress, David Hatcher. *Extraterrestrial Archaeology*. Revised 2nd Edition. Adventures Unlimited Press. 1995.

Collins, Andrew. *Gateway to Atlantis: The Search for the Source of a Lost Civilization.* Carroll and Graf Publishing. 2000.

Cook, Nick. *The Hunt for Zero Pointi.*

Corso, Col. Philip J. (Ret.), and Barnes, William J. *The Day After Roswell*. Pocket Books. 1997.
Cremo, Michael, and Thompson, Richard L. *Forbidden Archeology: The Hidden History of the Human Race.* Bhaktivedanta Book Publishing, Inc. 1996.
Davidson, Dan A. *Shape Power: A Treatise on How Form Converts Universal Aether into Electromagnetic and Gravitic Forces and Related Discoveries in Gravitational Physics.* Rivas Publishing. 1997.
Dingley, Gavin. "ParaSETI: ET Contact Via Subtle Energies," *Nexus,* Vol. 8, No. 1, January-Fedbruary 2001, pp. 37-43.
Dolan, *UFOs and the National Security State: An Unclassified History*, Volume One: *1941-1973*. Keyhole Publishing Company. 2000.
Dollard, Eric. "Transverse and Longitudinal Electric Waves." VHS Video. Borderland Sciences.
Dollard, Eric. *Condensed Introduction to Tesla Transformers.* Boderland Sciences. 1986.
Dollard, Eric. *Theory of Wireless Power.* Borderland Sciences. 1986.
Donato, William. "Cayce's Masers," *Atlantis Rising*, Number 32, March/April 2002, pp. 24-25, 61-62.
Ellis, Ralph. *Thoth: Architect of the Universe.* Adventures Unlimited Press. 2001.
Farrell, Joseph P. *The Giza Death Star*. Adventures Unlimited Press. 2002.
Flanagan, Dr. G. Patrick. *Pyramid Power: The Millenium Science.* Earthpulse Press. 1997.
Greer, Steven, M.D. *Disclosure: Military and Government Witnesses Reveal the Greatest Secrets in Modern History*. Crossing Point, Inc. 2001.
Hancock, Graham. *The Mars Mystery: The Secret Connection Between Earth and the Red Planet*. New York. Three Rivers Press. 1998.
Henshall, Philip. *The Nuclear Axis: Germany, Japan, and the Atom Bomb Race 1939-1945.* Sutton Publishing Limited. 2000.

Hessemann, Michael. *UFOs: The Secret History*. Marlowe and Company. 1998.

Hoagland, Richard C. *The Monuments of Mars: A City on the Edge of Forever.* North Atlantic Books. 1992.

Hoagland, Richard C. *The Monuments of Mars: A City on the Edge of Forever.* 5th Edition. North Atlantic Books. 2001.

Jochmans, Dr. Joseph, Litt D. "Hall of records: Opening soon?" *The Search for Lost Origins*. Atlantis Rising.1996.

Jochmans, Dr. Joseph, LittD. "How Old is the Great Pyramid, Really?" *The Search for Lost Origins*. Atlantis Rising. 1996.

Kenyon, J. Douglas. "Visitors from Beyond." *The Search for Lost Origins*. Atlantis Rising. 1996.

King, Moray B. *Quest for Zero Point energy: Engineering Principles for "Free Energy."* Adventures Unlimited Press. 2001.

Krasnoholovets, Dr. Volodymyr, "Sunmicroscopic Deterministic Quantum Mechanic."

Krasnoholovets, Volodymyr, Ph.D. "On the Way to Disclosing the Mysterious Power of the Great Pyramid."

Lemesurier, Peter. *The Great Pyramid Decoded.*

Mehler, Stephen S. *The Land of Osiris: An Introduction to Khemitology*. Adventures Unlimited Press. 2001.

Middleton-Jones, Howard, and Wilkie, James Michael. *Giza-Genesis: The Best Kept Secrets,* Vol. 1. Dandelion Books. 2001.

Pauwels, Louis, and Bergier, Michael. *The Morning of the Magicians*.

Penrose, Roger. *Shadows of the Mind: A Searching for the Missing Science of Consciousness.*

Picknett, Lynn, and Prince, Clive. *The Stargate Conspiracy*.

Possony, Dr. Stefan T. "The Tesla Connection." *Analysis of Scalar/Electromagnetic Technology*. Tesla Book Company, pp. 102-107.

Powers, Thomas, *Heisenberg's War: The Secret History of the German Bomb*. Da Capo Press. 1993.

Puthoff, Dr. Hal E., and Targ, Russell. "A Perceptual Channel for Information transfers over Kilometer Distances: Historical perspective and Recent Research."

Rubtsov, Dr. Vladimir V. "Domes of Wrath," *Fate*, April 2002, pp. 16-23.

Scranton, Laird. "The Dogon as Physicists," *Atlantis Rising*, Number 29, September/October 2001.

Swarz, Tim. *The Lost Journals of Nikola Tesla*. Global Communications. (No Date).

Temple, Robert. *The Sirius Mystery: New Scientific Evidence of Alien Contact 5,000 Years Ago*. Destiny Books. 1998.

Temple, Robert. *The Crystal Sun: Rediscovering a Lost technology of the Ancient World.* Century Books. 1999.

Tipler, Frank J. *The Physics of Immortality: Modern Cosmology, God, and the Resurrection of the Dead.* Doubleday. 1994.

Tompkins, Peter. *Secrets of the Great Pyramid.* New York. Harper and Row. 1971.

Torrun, Errol, "D&M Pyramid – Criteria." www.his.com/~tharsis/pyramid/criteria.

Van Flandern, Dr. Tom. *Dark Matter,, Missing Planets, and New Comets: Paradoxes Resolved, Origins Illuminated.* North Atlantic Books. 1993.

Vassilatos, Gerry. *Lost Science*. Adventures Unlimited Press. 1999.

Walker, Evan Harris. *The Physics of Consciousness: The Quantum Mind and the Meaning of Life*.

Whittaker, E.T. "An expression of the electromagnetic field due to electrons by means of two scalar potential functions," *Proceedings of the London Mathematcial Society*, Vol. 1, 1904, pp. 367-372.

Whittaker, E.T. "On the partial differential equations of mathematical physics," *Mathematische Analen*, Vol. 57, 1903, pp. 333-355.

Wilcox, Robert K. *Japan's Secret War: Japan's Race Against Time to Build its Own Atomic Bomb*. Marlowe and Company. 1995.

Books Consulted for, or Cited in, *The Giza Death Star Destroyed:*

Alouf, M. *The History of Baalbek.*

Boulay, R.A. *Flying Serpents and Dragons: The Story of Mankind's Reptilian Past*. Revised and Expanded Edition. Book Tree. 2001.

Bounias, Michael, and Krasnoholovets, Volydymyr. "Scanning the Structure of Ill-Known Spaces, Pt. 1: Founding Principles About Mathematical Constitutions of Space." *Kybernetics: The International Journal for Systems and Cybernetics*. 2002.

Bounias, Michael, and Krasnoholovets, Volydymyr. "Scanning the Structure of Ill-Known Spaces, Pt. 2: Principles of Construction of Physical Space." *Kybernetics: The International Journal for Systems and Cybernetics*. 2002.

Brennan, Herbie. *The Secret History of Ancient Egypt.* Berkeley Books. 2000.

De Grazia, Alfred, ed. *The Velikhovsky Affair: The Warfare of Science and Scientism* University Books. 1996.

DeSalvo, John, Ph.D. *The Complete Pyramid Sourcebook.* No Place, No Publisher. 2003. ISBN 1-4107-8041-0.

Enuma Elish, trans. and ed. L.W. Kng, M.A. F.S.A., Vol 1, Luzac and Company. 1902.

Farrell, Joseph P. *Free Choice in St. Maximus the Confessor*. St. Tikhon's Seminary Press. 1990.

Farrell, Joseph P. *The Giza Death Star*. Adventures Unlimited Press. 2001.

Farrell, Joseph P. *The Giza Death Star Deployed.* Adventures Unlimited Press. 2003.

Farrell, Joseph P. *Reich of the Black Sun: Nazi Secret Weapons and the Cold War Allied Legend.* Adventures Unlimited Press. 2004.

Flynn, David. *Cydonia: The Secret Chronicles of Mars*. End Time Thunder Publishers. 2002.

Gardner, Laurence. *Genesis of the Grail Kings: The Explosive*

Story of Genetic Cloning and the Ancient Bloodline of Jesus. Fair Winds. 2001.

Gardner, Laurence. *Lost Secrets of the Sacred Ark.*

Hall, Manly P. *The Secret Teachings of All Ages: AN Encyclopedic Outline of Masonic, Hermetic, Qabbalistic, and Rosicrucian Symbolical Philosophy.* Reader's Edition.

Knight, Christopher, and Lomas, Robert. *Uriel's Machine: Uncovering the Secrets of Stonehenge, Noah's Flood, and the Dawn of Civilization.* Barnes and Noble. 2004.

LaViolette, Paul A. *Subquantum Kinetics: The Alchemy of Creation*. 1994.

Lomas, Robert. *Freemasonry and the Rise of Modern Science.* Fair Winds Press. 2004.

McCanney, James. *Planet-X, Comets, and Earth Changes: A Scientific Treatise on the Effects of a New Large Planet of Comet arriving in our Solar System and Expected Weather and Earth Changes.*

Moody, Richard Jr. "Albert Einstein: Plagiarist of the Century," *Nexus* Vol. 11, No. 1, January-February 2004, pp. 43-46, 74.

Mortobn, Chris, and Thomas, Ceri Louise. *The Mystery of the Crystal Skulls: Unlocking the Secrets of the Past, present, and Future.*

Müller, dr. hartmut, "An Introduction to Global Scaling Theory," *Nexus*, Vol. 11, No. 5, September—October 2004.

Plotinus, *Enneads*. Trans H.E. Armstrong.

Rux, Bruce. *Architects of the Underworld: Unriddling Atlantis, Anomalies of Mars, and the Mystery of the Sphinx*. Frog, Ltd. 1996.

Schwaller de Lubicz, R.A. *The Temple of Man: Aept of the South at Luxor.* Trans. from the French by Deborah Lawlor and Roboert Lawlor. Inner Traditions. 1998.

Scott, Sir Walter. Trans. and ed. *Hermetica*. Shambala.

Van Flandern, Dr. Tom. *Dark Matter, Missing Planets, and New Comets: Paradoxes Resolved, Origins Illuminated.* North Atlantic Books. 1993.

West, John Anthony. *The Serpent in the Sky: The High Wisdom of Ancient Egypt*. New York. Harper and Row. 1999.

Wikipedia. www.wikipedia.org/wiki/Sumerian_king_list

Wilkins, Harold T. *Mysteries of Ancient South America.* Adventures Unlimited Press.

Books Consulted for, or Cited in, This Work

------. "Acoustic metamaterial." *Wikipedia.* https:// en.wikipedia.org/wiki/Acoustic_metamaterial.

-----. "Optical Rotation." *Wikipedia.* https://en.wikipedia.org/ wiki/ Optical_rotation

Addazi, Adnrea; Marciano, Antonino; Pasechnik, Roman. "Time-crystal ground state and production of gravitational waves from QCD phase transition." *Chinese Physics C*. Vol. 43, No. 6, 2019, pp. 065101-1 –065101-11.

Alu, Andrea. "Engineered Metamaterials Can Trick Light and Sound into Mind-Bending Behavior: Advanced Materials can modify waves, creating optical illusions and useful technologies." *Scientific American.* November, 2022.

American Physical Society. "Gravity crystals: A new method for exploring the physics of white dwarf stars." *Phys.Org.* October 21, 2019.

Blair, David. "A tiny crystal device could boost gravitational wave detectors to reveal the birth cries of black holes." *Phys.Org.* February 16, 2021.

Chandler, David L. "Trapping light with a twister: New understanding of how to halt photons could lead to miniature particle accelerators, improved data transmission. *MIT News*. https://news.mit.edu/2014/ trapping-light-minature-particle-accelrators-omproved-data-transmission-1222.

Chen, Mingzhou, Mazilu, Michael, Arita, Yoshihiko, Wright, Ewan M., and Dholakia, Kishan. "Dynamics of microparticles trapped in a perfect vortex beam." *Optics Letters*, vol. 38, Issue 22, 2013. pp. 4919-4922.

Clarkeson, Antonia. "Explaining Newton's *Dissertation upon the Sacred Cubit of the Jews*". http://users/tpg.com/au/ adsley22 /FHG/9%20Newton's%20Dissertation.pdf.

Chreighton, Scott. *The Great Pyramid Hoax: The Conspiracy to Conceal the True History of Ancient Egypt*. Rochester, Vermont. Bear and Company. 2017. ISBN 978-1-59143-789-5

Cross, Spencer L. *The Great Pyramid: A Factory for Mono-Atomic Gold*. No publisher information. 2013. ISBN 978-0615919768.

Geller, Samuel. *Die Sumerisch-Assyrische Serie Lugal-E ud Me-lam-bi Nir-Gal*. London. Forgotten Books. 2018. ISBN978-1-390-35728-8.

Hazen, Robert M. *The New Alchemists: Breaking through the Barriers of High Pressure*. New York. Random House Times Books. 1993. ISBN 0-8129-2275-1.

Hutchings, N.W. *The Great Pyramid: Prophecy in Stone*. Crane, Missouri. Defender Publishing. '

Jessup, M.K. *The Case for the UFO: Unidentified Flying Objects, The Varo Edition*. No date or place of publication. The Anomalies Network. ISBN 978-1479151431.

Kastalia Medrano. "Physics: Speed of Light could be Brought to a Complete Stop by Trapping Particles Inside Crystals." January 2, 2018.

Kingsland, William. *The Great Pyramid in Fact and Theory*. No Date. Literary Licensing. ISBN 978149791417.

Kunz, George Frederick. *The Curious Lore of Precious Stones*. East Mineola, New York. Dover Publications.1971. ISBN 0-486-22227-6.

Liang, Yansheng; Lei, Ming; Yan, Shaohui; Li, Manman; Cai, Yanan; Wang; Zhaojun, Yu, Zianghua; Yao,Baoli. "Rotating of low-refractice-index microparticles with a quasi-perfect optical vortex." *Applied Optics*, January 1:57(1), pp. 79-84, 2018.

Meissner, Bruno. *Altorientalische texte und Untersuchungen*. Longon. Forgotten Books. 2018. ISBN 978-1-332-62308-2.

Members of the David H. Kkoch Pyramids Radiocarbon Project. "Dating the Pyramids," *Archaeology*. Volume 52, Number 5, September/October 1999.

Newton, Sir Isaac. *A Dissertation upon the Sacred Cubit of the Jews and the Cunits of the Several Nations*. https://newtonproject. ox.ac.uk/view/texts.normalized/THEM00276.

Schmitz, Eckhart R. *The Great Pyramid of Giza: Decoding the Measure of a Monument.* Nepean, Ontario. Roland Publishing. 2012. ISBN 978-0-9879577-0-2.

Skinner, J. Ralston. *Key to the Hebrew-Egyptian Mystery in the Source of Measures Originating the British Inch.* No place or date. Wentworth Press. ISBN 978-0526709991.

Taylor, John. *The Great Pyramid: Why was it Built? And Who Built It?* Cambridge University Press. 2014. ISBN 978-1-108-07578-7.

Van Auken, John, *2038: The Great Pyramid Timeline Prophecy.* Virginia Beach, Virginia. A.R.E. Press. 2012.

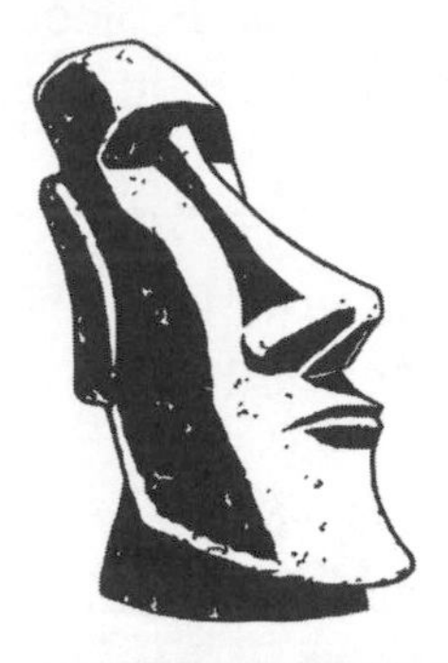

Get these fascinating books from your nearest bookstore or directly from: Adventures Unlimited Press

www.adventuresunlimitedpress.com

COVERT WARS AND BREAKAWAY CIVILIZATIONS
By Joseph P. Farrell

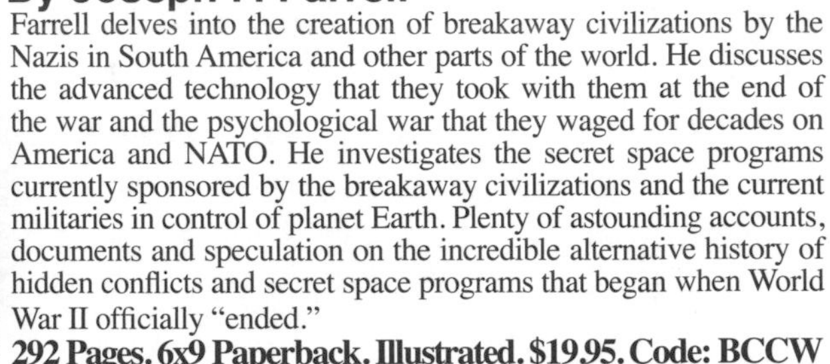

Farrell delves into the creation of breakaway civilizations by the Nazis in South America and other parts of the world. He discusses the advanced technology that they took with them at the end of the war and the psychological war that they waged for decades on America and NATO. He investigates the secret space programs currently sponsored by the breakaway civilizations and the current militaries in control of planet Earth. Plenty of astounding accounts, documents and speculation on the incredible alternative history of hidden conflicts and secret space programs that began when World War II officially "ended."

292 Pages. 6x9 Paperback. Illustrated. $19.95. Code: BCCW

THE ENIGMA OF CRANIAL DEFORMATION
Elongated Skulls of the Ancients
By David Hatcher Childress and Brien Foerster

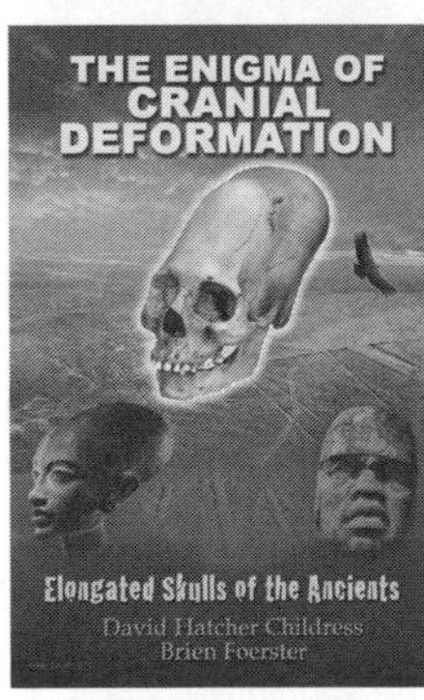

In a book filled with over a hundred astonishing photos and a color photo section, Childress and Foerster take us to Peru, Bolivia, Egypt, Malta, China, Mexico and other places in search of strange elongated skulls and other cranial deformation. The puzzle of why diverse ancient people—even on remote Pacific Islands—would use head-binding to create elongated heads is mystifying. Where did they even get this idea? Did some people naturally look this way—with long narrow heads? Were they some alien race? Were they an elite race that roamed the entire planet? Why do anthropologists rarely talk about cranial deformation and know so little about it? Color Section.

250 Pages. 6x9 Paperback. Illustrated. $19.95. Code: ECD

ARK OF GOD
The Incredible Power of the Ark of the Covenant
By David Hatcher Childress

Childress takes us on an incredible journey in search of the truth about (and science behind) the fantastic biblical artifact known as the Ark of the Covenant. This object made by Moses at Mount Sinai—part wooden-metal box and part golden statue—had the power to create "lightning" to kill people, and also to fly and lead people through the wilderness. The Ark of the Covenant suddenly disappears from the Bible record and what happened to it is not mentioned. Was it hidden in the underground passages of King Solomon's temple and later discovered by the Knights Templar? Was it taken through Egypt to Ethiopia as many Coptic Christians believe? Childress looks into hidden history, astonishing ancient technology, and a 3,000-year-old mystery that continues to fascinate millions of people today. Color section.

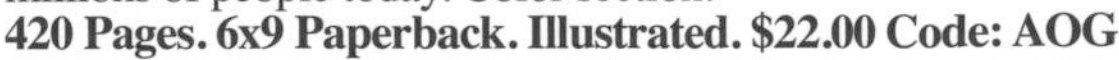

420 Pages. 6x9 Paperback. Illustrated. $22.00 Code: AOG

HESS AND THE PENGUINS
The Holocaust, Antarctica and the Strange Case of Rudolf Hess
By Joseph P. Farrell

Farrell looks at Hess' mission to make peace with Britain and get rid of Hitler—even a plot to fly Hitler to Britain for capture! How much did Göring and Hitler know of Rudolf Hess' subversive plot, and what happened to Hess? Why was a doppleganger put in Spandau Prison and then "suicided"? Did the British use an early form of mind control on Hess' double? John Foster Dulles of the OSS and CIA suspected as much. Farrell also uncovers the strange death of Admiral Richard Byrd's son in 1988, about the same time of the death of Hess.

288 Pages. 6x9 Paperback. Illustrated. $19.95. Code: HAPG

HIDDEN FINANCE, ROGUE NETWORKS & SECRET SORCERY
The Fascist International, 9/11, & Penetrated Operations
By Joseph P. Farrell

Farrell investigates the theory that there were not *two* levels to the 9/11 event, but *three*. He says that the twin towers were downed by the force of an exotic energy weapon, one similar to the Tesla energy weapon suggested by Dr. Judy Wood, and ties together the tangled web of missing money, secret technology and involvement of portions of the Saudi royal family. Farrell unravels the many layers behind the 9-11 attack, layers that include the Deutschebank, the Bush family, the German industrialist Carl Duisberg, Saudi Arabian princes and the energy weapons developed by Tesla before WWII.

296 Pages. 6x9 Paperback. Illustrated. $19.95. Code: HFRN

THRICE GREAT HERMETICA & THE JANUS AGE
By Joseph P. Farrell

What do the Fourth Crusade, the exploration of the New World, secret excavations of the Holy Land, and the pontificate of Innocent the Third all have in common? Answer: Venice and the Templars. What do they have in common with Jesus, Gottfried Leibniz, Sir Isaac Newton, Rene Descartes, and the Earl of Oxford? Answer: Egypt and a body of doctrine known as Hermeticism. The hidden role of Venice and Hermeticism reached far and wide, into the plays of Shakespeare (a.k.a. Edward DeVere, Earl of Oxford), into the quest of the three great mathematicians of the Early Enlightenment for a lost form of analysis, and back into the end of the classical era, to little known Egyptian influences at work during the time of Jesus.

354 Pages. 6x9 Paperback. Illustrated. $19.95. Code: TGHJ

REICH OF THE BLACK SUN
Nazi Secret Weapons & the Cold War Allied Legend
by Joseph P. Farrell

Why were the Allies worried about an atom bomb attack by the Germans in 1944? Why did the Soviets threaten to use poison gas against the Germans? Why did Hitler in 1945 insist that holding Prague could win the war for the Third Reich? Why did US General George Patton's Third Army race for the Skoda works at Pilsen in Czechoslovakia instead of Berlin? Why did the US Army not test the uranium atom bomb it dropped on Hiroshima? Why did the Luftwaffe fly a non-stop round trip mission to within twenty miles of New York City in 1944? Farrel takes the reader on a scientific-historical journey in order to answer these questions. Arguing that Nazi Germany won the race for the atom bomb in late 1944,

352 PAGES. 6x9 PAPERBACK. ILLUSTRATED. $16.95. CODE: ROBS

THE COSMIC WAR
Interplanetary Warfare, Modern Physics, and Ancient Texts
By Joseph P. Farrell

There is ample evidence across our solar system of catastrophic events. The asteroid belt may be the remains of an exploded planet! The known planets are scarred from incredible impacts, and teeter in their orbits due to causes heretofore inadequately explained. Included: The history of the Exploded Planet hypothesis, and what mechanism can actually explode a planet. The role of plasma cosmology, plasma physics and scalar physics. The ancient texts telling of such destructions: from Sumeria (Tiamat's destruction by Marduk), Egypt (Edfu and the Mars connections), Greece (Saturn's role in the War of the Titans) and the ancient Americas.

436 Pages. 6x9 Paperback. Illustrated.. $18.95. Code: COSW

THE GRID OF THE GODS
The Aftermath of the Cosmic War & the Physics of the Pyramid Peoples
By Joseph P. Farrell with Scott D. de Hart

Farrell looks at Ashlars and Engineering; Anomalies at the Temples of Angkor; The Ancient Prime Meridian: Giza; Transmitters, Nazis and Geomancy; the Lithium-7 Mystery; Nazi Transmitters and the Earth Grid; The Master Plan of a Hidden Elite; Moving and Immoveable Stones; Uncountable Stones and Stones of the Giants and Gods; The Grid and the Ancient Elite; Finding the Center of the Land; The Ancient Catastrophe, the Very High Civilization, and the Post-Catastrophe Elite; Tiahuanaco and the Puma Punkhu Paradox: Ancient Machining; The Black Brotherhood and Blood Sacrifices; The Gears of Giza: the Center of the Machine; tons more.

436 Pages. 6x9 Paperback. Illustrated. $19.95. Code: GOG

THE SS BROTHERHOOD OF THE BELL
The Nazis' Incredible Secret Technology
by Joseph P. Farrell

In 1945, a mysterious Nazi secret weapons project code-named "The Bell" left its underground bunker in lower Silesia, along with all its project documentation, and a four-star SS general named Hans Kammler. Taken aboard a massive six engine Junkers 390 ultra-long range aircraft, "The Bell," Kammler, and all project records disappeared completely, along with the gigantic aircraft. It is thought to have flown to America or Argentina. What was "The Bell"? What new physics might the Nazis have discovered with it? How far did the Nazis go after the war to protect the advanced energy technology that it represented?

456 pages. 6x9 Paperback. Illustrated. $16.95. Code: SSBB

THE THIRD WAY
The Nazi International, European Union, & Corporate Fascism
By Joseph P. Farrell

Pursuing his investigations of high financial fraud, international banking, hidden systems of finance, black budgets and breakaway civilizations, Farrell continues his examination of the post-war Nazi International, an "extra-territorial state" without borders or capitals, a network of terrorists, drug runners, and people in the very heights of financial power willing to commit financial fraud in amounts totaling trillions of dollars. Breakaway civilizations, black budgets, secret technology, international terrorism, giant corporate cartels, patent law and the hijacking of nature: Farrell explores 'the business model' of the post-war Axis elite.

364 Pages. 6x9 Paperback. Illustrated. $19.95. Code: TTW

ROSWELL AND THE REICH
The Nazi Connection
By Joseph P. Farrell

Farrell has meticulously reviewed the best-known Roswell research from UFO-ET advocates and skeptics alike, as well as some little-known source material, and comes to a radically different scenario of what happened in Roswell, New Mexico in July 1947, and why the US military has continued to cover it up to this day. Farrell presents a fascinating case sure to disturb both ET believers and disbelievers, namely, that what crashed may have been representative of an independent postwar Nazi power—an extraterritorial Reich monitoring its old enemy, America, and the continuing development of the very technologies confiscated from Germany at the end of the War.

540 pages. 6x9 Paperback. Illustrated. $19.95. Code: RWR

SECRETS OF THE UNIFIED FIELD
The Philadelphia Experiment, the Nazi Bell, and the Discarded Theory
by Joseph P. Farrell

Farrell examines the now discarded Unified Field Theory. American and German wartime scientists and engineers determined that, while the theory was incomplete, it could nevertheless be engineered. Chapters include: The Meanings of "Torsion"; Wringing an Aluminum Can; The Mistake in Unified Field Theories and Their Discarding by Contemporary Physics; Three Routes to the Doomsday Weapon: Quantum Potential, Torsion, and Vortices; Tesla's Meeting with FDR; Arnold Sommerfeld and Electromagnetic Radar Stealth; Electromagnetic Phase Conjugations, Phase Conjugate Mirrors, and Templates; The Unified Field Theory, the Torsion Tensor, and Igor Witkowski's Idea of the Plasma Focus; tons more.

340 pages. 6x9 Paperback. Illustrated. $18.95. Code: SOUF

NAZI INTERNATIONAL
The Nazi's Postwar Plan to Control Finance, Conflict, Physics and Space
by Joseph P. Farrell

Beginning with prewar corporate partnerships in the USA, including some with the Bush family, he moves on to the surrender of Nazi Germany, and evacuation plans of the Germans. He then covers the vast, and still-little-known recreation of Nazi Germany in South America with help of Juan Peron, I.G. Farben and Martin Bormann. Farrell then covers Nazi Germany's penetration of the Muslim world including Wilhelm Voss and Otto Skorzeny in Gamel Abdul Nasser's Egypt before moving on to the development and control of new energy technologies including the Bariloche Fusion Project, Dr. Philo Farnsworth's Plasmator, and the work of Dr. Nikolai Kozyrev. Finally, Farrell discusses the Nazi desire to control space, and examines their connection with NASA, the esoteric meaning of NASA Mission Patches.

412 pages. 6x9 Paperback. Illustrated. $19.95. Code: NZIN

ARKTOS
The Polar Myth in Science, Symbolism & Nazi Survival
by Joscelyn Godwin

Explored are the many tales of an ancient race said to have lived in the Arctic regions, such as Thule and Hyperborea. Progressing onward, he looks at modern polar legends: including the survival of Hitler, German bases in Antarctica, UFOs, the hollow earth, and the hidden kingdoms of Agartha and Shambala. Chapters include: Prologue in Hyperborea; The Golden Age; The Northern Lights; The Arctic Homeland; The Aryan Myth; The Thule Society; The Black Order; The Hidden Lands; Agartha and the Polaires; Shambhala; The Hole at the Pole; Antarctica; more.

220 Pages. 6x9 Paperback. Illustrated. Bib. Index. $16.95. Code: ARK

HAUNEBU: THE SECRET FILES
The Greatest UFO Secret of All Time
By David Hatcher Childress

Childress brings us the incredible tale of the German flying disk known as the Haunebu. Although rumors of German flying disks have been around since the late years of WWII it was not until 1989 when a German researcher named Ralf Ettl living in London received an anonymous packet of photographs and documents concerning the planning and development of at least three types of unusual craft. Chapters include: A Saucer Full of Secrets; WWII as an Oil War; A Saucer Called Vril; Secret Cities of the Black Sun; The Strange World of Miguel Serrano; Set the Controls for the Heart of the Sun; Dark Side of the Moon: more. Includes a 16-page color section. Over 120 photographs and diagrams.

352 Pages. 6x9 Paperback. Illustrated. $22.00 Code: HBU

ANTARCTICA AND THE SECRET SPACE PROGRAM
Hatcher Childress

David Childress, popular author and star of the History Channel's show *Ancient Aliens*, brings us the incredible tale of Nazi submarines and secret weapons in Antarctica and elsewhere. He then examines Operation High-Jump with Admiral Richard Byrd in 1947 and the battle that he apparently had in Antarctica with flying saucers. Through "Operation Paperclip," the Nazis infiltrated aerospace companies, banking, media, and the US government, including NASA and the CIA after WWII. Does the US Navy have a secret space program that includes huge ships and hundreds of astronauts?

392 Pages. 6x9 Paperback. Illustrated. $22.00 Code: ASSP

THE ANTI-GRAVITY FILES
A Compilation of Patents and Reports
Edited by David Hatcher Childress

With plenty of technical drawings and explanations, this book reveals suppressed technology that will change the world in ways we can only dream of. Chapters include: A Brief History of Anti-Gravity Patents; The Motionless Electromagnet Generator Patent; Mercury Anti-Gravity Gyros; The Tesla Pyramid Engine; Anti-Gravity Propulsion Dynamics; The Machines in Flight; More Anti-Gravity Patents; Death Rays Anyone?; The Unified Field Theory of Gravity; and tons more. Heavily illustrated. 4-page color section.

216 pages. 8x10 Paperback. Illustrated. $22.00. Code: AGF

SECRET MARS: The Alien Connection
By M. J. Craig

While scientists spend billions of dollars confirming that microbes live in the Martian soil, people sitting at home on their computers studying the Mars images are making far more astounding discoveries… they have found the possible archaeological remains of an extraterrestrial civilization. Hard to believe? Well, this challenging book invites you to take a look at the astounding pictures yourself and make up your own mind. *Secret Mars* presents over 160 incredible images taken by American and European spacecraft that reveal possible evidence of a civilization that once lived, and may still live, on the planet Mars… powerful evidence that scientists are ignoring! A visual and fascinating book!

352 Pages. 6x9 Paperback. Illustrated. $19.95. Code: SMAR

ORDER FORM

10% Discount When You Order 3 or More Items!

One Adventure Place
P.O. Box 74
Kempton, Illinois 60946
United States of America
Tel.: 815-253-6390 • Fax: 815-253-6300
Email: auphq@frontiernet.net
http://www.adventuresunlimitedpress.com

ORDERING INSTRUCTIONS

- ✓ Remit by USD$ Check, Money Order or Credit Card
- ✓ Visa, Master Card, Discover & AmEx Accepted
- ✓ Paypal Payments Can Be Made To: info@wexclub.com
- ✓ Prices May Change Without Notice
- ✓ 10% Discount for 3 or More Items

SHIPPING CHARGES

United States

- ✓ POSTAL BOOK RATE
- ✓ Postal Book Rate { $5.00 First Item; 50¢ Each Additional Item
- ✓ Priority Mail { $8.50 First Item; $2.00 Each Additional Item
- ✓ UPS { $9.00 First Item (Minimum 5 Books); $1.50 Each Additional Item

NOTE: UPS Delivery Available to Mainland USA Only

Canada

- ✓ Postal Air Mail { $19.00 First Item; $3.00 Each Additional Item
- ✓ Personal Checks or Bank Drafts MUST BE US$ and Drawn on a US Bank
- ✓ Canadian Postal Money Orders OK
- ✓ Payment MUST BE US$

All Other Countries

- ✓ Sorry, No Surface Delivery!
- ✓ Postal Air Mail { $29.00 First Item; $7.00 Each Additional Item
- ✓ Checks and Money Orders MUST BE US$ and Drawn on a US Bank or branch.
- ✓ Paypal Payments Can Be Made in US$ To: info@wexclub.com

SPECIAL NOTES

- ✓ RETAILERS: Standard Discounts Available
- ✓ BACKORDERS: We Backorder all Out-of-Stock Items Unless Otherwise Requested
- ✓ PRO FORMA INVOICES: Available on Request
- ✓ DVD Return Policy: Replace defective DVDs only

ORDER ONLINE AT: www.adventuresunlimitedpress.com

10% Discount When You Order 3 or More Items!

Please check: ✓

☐ This is my first order ☐ I have ordered before

Name

Address

City

State/Province | Postal Code

Country

Phone: Day | Evening

Fax | Email

Item Code	Item Description	Qty	Total

Please check: ✓

- ☐ Postal-Surface
- ☐ Postal-Air Mail (Priority in USA)
- ☐ UPS (Mainland USA only)

Subtotal ▸
Less Discount-10% for 3 or more items ▸
Balance ▸
Illinois Residents 6.25% Sales Tax ▸
Previous Credit ▸
Shipping ▸
Total (check/MO in USD$ only) ▸

☐ Visa/MasterCard/Discover/American Express

Card Number:

Expiration Date: | Security Code:

✓ SEND A CATALOG TO A FRIEND: